Springer

Berlin
Heidelberg
New York
Hong Kong
London
Milan
Paris
Tokyo

F. H. Gage   A. Björklund
A. Prochiantz   Y. Christen (Eds.)

Stem Cells
in the Nervous System:
Functional and Clinical
Implications

With 29 Figures and 3 Tables

 Springer

Gage, Fred H., Ph.D.
The Salk Institute for
Biological Studies
10010 N. Torrey Pines Road
La Jolla, CA 92037
USA
e-mail: gage@salk.edu

Prochiantz, Alain, Ph.D.
Ecole Normale Supérieure and
CNRS UMP8542
46 rue d'Ulm
75005 Paris
France
e-mail: prochian@wotan.ens.fr

Björklund, Anders, M.D.
Wallenberg Neurosciences Center
Department of Physiological Sciences
Lund University, BMC A11
22184 Lund
Sweden
e-mail: anders.bjorklund@mphy.lu.se

Christen, Yves, Ph.D.
Fondation IPSEN
Pour la Recherche Thérapeutique
24, rue Erlanger
75781 Paris Cedex 16
France
e-mail: yves.christen@ipsen.com

ISBN 3-540-20558-6   Springer-Verlag Berlin Heidelberg New York

Cataloging-in-Publication Data applied for Bibliographic information published by Die Deutsche Bibliothek
Die Deutsche Bibliothek lists this publication in the Deutsche Nationalbibliografie; detailed bibliographic data is available in the Internet at <http://dnb.ddb.de>.

Springer-Verlag is a part of Springer Science+Business Media

springeronline.com

© Springer-Verlag Berlin Heidelberg 2004
Printed in Germany

The use of general descriptive names, registered names, trademarks, etc. in this publications does not imply, even in the absence of a specific statement, that such names are exempt from the relevant protective laws and regulations and therefore free for general use.

Product liability: The publishers cannot guarantee the accuracy of any information about dosage and application contained in this book. In every individual case the user must check such information by consulting the relevant literature.

Production: PRO EDIT GmbH, 69126 Heidelberg, Germany
Cover design: design & production, 69121 Heidelberg, Germany
Typesetting: Satz & Druckservice, 69181 Leimen, Germany
Printed on acid-free paper   27/3150Re   5 4 3 2 1 0

Preface

On September 18, 1995, Fondation Ipsen devoted one of its *Colloque Médecine et Recherche* to stem cells and how they can be used to understand and, possibly, treat the nervous system. At the time, this was one of the very first international meetings on this topic. Since then, major developments in the area have been unceasing, whether in terms of basic science or the prospects that such research offers. The said prospects have given rise to a great deal of hope and, sometimes, concern, although it should be said that the latter resulted from oversimplified approaches to the matter, or approaches based more on science-fiction than on biological science. In reality, research on neurogenesis in the central nervous system has yielded both highly interesting data and a tremendous amount of questions and uncertainties, which require, above all else, serious scientific data and new experiments be set forth.

In the scientific realm, discussions have significantly changed in focus since 1995, and now deal less with the existence of stem cells in the adult brain, than with their functional integration and the part they play in neurophysiological processes such as learning, or following cerebral lesions. In order to help further research in this area, Fondation Ipsen decided to return to the topic of stem cells for its 12th *Colloque Médecine et Recherche* in Neurosciences. The meeting was held on 20 January 2003, in Paris. This volume presents the proceedings of this symposium.

At the meeting, Fondation Ipsen's Neuronal Plasticity Prize was awarded by the President of the Jury, Pr Jean-Pierre Changeux, to three research pioneers in this field: Arturo Alvarez-Buylla, Ronald McKay and Samuel Weiss. Other precursors, Fred Gage and Anders Björklund, had previously received the same distinction, in 1990. All of the aforementioned researchers took part in the meeting and in this book.

January 2004   *Yves Christen*

Acknowledgment
The editors want to thank J. Mervaillie for the organization of the meeting and M.-L. Gage for the editing of the book.

Contents

Contributors

Alvarez-Buylla, Arturo
Department of Neurological Surgery, Brain Tumor Research Center, Box 0520,
533 Parnassus Ave, San Francisco, CA 94143-0520, USA

Anderson, David, J.
Division of Biology 216-76, Howard Hughes Medical Institute,
California Institute of Technology, Pasadena, CA 91125, USA

Arvidsson, Andreas
Section of Restorative Neurology, Wallenberg Neuroscience Center,
University Hospital BMC A11, 221 84 Lund, Sweden

Björklund, Anders
Wallenberg Neuroscience Center, Dept. of Physiological Sciences,
Lund University, BMC A11, 221 84 Lund, Sweden

Caillé, Isabelle
Ecole Normale Supérieure and CNRS UMR 8542, 46 rue d'Ulm,
75005 Paris, France

Choi, Gloria
Division of Biology 216-76, Howard Hughes Medical Institute,
California Institute of Technology, Pasadena, CA 91125, USA

Doetsch, Fiona
Department of Neurological Surgery, Brain Tumor Research Center,
Box 0520, 533 Parnassus Ave., San Francisco, CA 94143-0520, USA

Ekdahl, Christine
Section of Restorative Neurology, Wallenberg Neuroscience Center,
University Hospital BMC A11, 221 84 Lund, Sweden

Englund, Ulrica
Department of Neurodegenerative Disorders, H. Lundbeck A/S,
Ottiliavej 9, 2500 Valby, Denmark

Enwere, Emeka
Genes and Development Research Group, University of Calgary
Faculty of Medicine, Calgary, Alberta, Canada T2N4N1

Frisén, Jonas
Department of Cell and Molecular Biology, Medical Nobel Institute,
Karolinska Institute, 17177 Stockholm, Sweden

Gage, Fred, H.
The Salk Institute for Biological Studies,
10010 N. Torrey Pines Road, La Jolla, CA 92037, USA

Garcia-Verdugo, José Manuel
Department of Neurological Surgery, Brain Tumor Research Center,
Box 0520, 533 Parnassus Ave., San Francisco, CA 94143-0520, USA

Jamieson, Catriona H.M.
Division of Hematology, Department of Medicine,
Stanford University School of Medicine, and Department of Pathology,
Stanford University School of Medicine B257 Beckman Center,
279 Campus Drive, Stanford, CA 94305-5323, USA

Joliot, Alain
Ecole Normale Supérieure and CNRS UMR8542,
46 rue d'Ulm, 75005 Paris, France

Kempermann, Gerd
Max Delbrück Center for Molecular Medicine Berlin-Buch,
Robert-Rössle-Str. 10, 13125 Berlin and
Volkswagen Foundation Research Group, Dept. of Experimental Neurology,
Humboldt University Berlin, Germany

Kokaia, Zaal
Section of Restorative Neurology, Wallenberg Neuroscience Center,
University Hospital BMC A11, 221 84 Lund, Sweden

Lesaffre, Brigitte
Ecole Normale Supérieure and CNRS UMR 8542,
46 rue d'Ulm, 75005 Paris, France

Lindvall, Olle
Section of Restorative Neurology, Wallenberg Neuroscience Center,
University Hospital BMC A11, 221 84 Lund, Sweden

Lo, Liching
Division of Biology 216-76, Howard Hughes Medical Institute,
California Institute of Technology, Pasadena, CA 91125, USA

Mainguy, Gaëll
Ecole Normale Supérieure and CNRS UMR 8542,
46 rue d'Ulm, 75005 Paris, France

Mc Kay, Ronald
Laboratory of Molecular Biology, NIDS, NIH,
Bldg. 36 Rm 5A29, Convent Dr-MSC 4157, Bethesda, MD 20892-4157, USA

Passegué, Emmanuelle
Department of Pathology, Stanford University School of Medicine,
B257 Beckman Center, 279 Campus Drive, Stanford, CA 94305-5323, USA

Prochiantz, Alain
Ecole Normale Supérieure and CNRS UMR 8542,
46 rue d'Ulm, 75005 Paris, France

Schaffer, David
Department of Chemical Engineering, University of California,
Berkeley, CA 94720, USA

Seri, Bettina
Department of Neurological Surgery, Brain Tumor Research Center,
Box 0520, 533 Parnassus Ave., San Francisco, CA 94143-0520, USA

Smith, Austin
Institute for Stem Cell Research, University of Edinburgh,
King's Buildings, West Mains Road, Edinburgh, EH9 3JQ, Scotland

Sonnier, Laure
Ecole Normale Supérieure and CNRS UMR 8542,
46 rue d'Ulm, 75005 Paris, France

Volovitch, Michel
Ecole Normale Supérieure and CNRS UMR 8542,
46 rue d'Ulm, 75005 Paris, France

Weiss, Samuel
Genes and Development Research Group, University of Calgary
Faculty of Medicine, Calgary, Alberta, Canada T2N4N1

Weissman, Irving L.
Department of Pathology, Stanford University School of Medicine,
B257 Beckman Center, 279 Campus Drive, Stanford, CA 94305-5323, USA

Wiskott, Laurenz
Volkswagen Foundation Research Group, Institute for Theoretical Biology,
Humboldt University Berlin, Invalidenstrasse 43,10115 Berlin, Germany

Zhao, Xinyu
The Salk Institute for Biological Studies,
10010 N. Torrey Pines Road, La Jolla, CA 92037, USA

Zhou, Qiao
Division of Biology 216-76, Howard Hughes Medical Institute,
California Institute of Technology, Pasadena, CA 91125, USA

Zirlinger, Mariela
Department of Biochemistry and Molecular Biology, Harvard University,
16 Divinity Ave, Cambridge, MA 02138, USA

Neurogenesis in adult brain: understanding its mechanism and regulation

X. Zhao[1], D. Schaffer[2], and F.H. Gage[3]

Adult neurogenesis is an intriguing phenomenon because of the promise it holds for leading to a new understanding of brain functions and in providing new therapeutic tools for regeneration and repair of diseased and injured adult CNS. We need to truly understand adult neurogenesis – what triggers it, what inhibits it and how it is regulated – before we can use this phenomenon appropriately.

Is adult neurogenesis functional?

After 40 years of research, scientists have finally confirmed that persistent neurogenesis occurs at least in the hippocampus and subventricular zone (SVZ) of the adult mammalian brain (Altman 1965; Gould et al. 1999; Kaplan and Hinds 1977; Kempermann et al. 1997b; Eriksson et al. 1998; Gage 2002; Kornack and Rakic 1999). The obvious next question is: "Are the newly generated neurons functional?" If so, "What are the functions of these new neurons?"

The first question has been answered by a few critical experiments. Cultured adult neural progenitor cells (NPCs) have been shown to differentiate into cells that meet all the criteria of being neurons, including having neuronal polarity, expressing neuron-specific markers, forming functional synapses, being able to elicit tetrodotoxin-sensitive action potentials, and being able to communicate with other neurons by releasing and detecting neurotransmitters at their synapses (Song et al. 2002b). Such results demonstrate that adult NPCs can make functional neurons at least in vitro when they are co-cultured with primary hippocampal astrocytes. Another advance came with a study showing functional neurogenesis in the adult hippocampus, (van Praag et al. 2002) injected mice with a GFP-expressing retrovirus that infects only proliferating cells and can be visualized in live hippocampal slices. They demonstrated that new granule neurons in the hippocampus exhibited neuronal morphology and displayed passive membrane potentials, action potentials and synaptic inputs that were similar to mature dentate neurons (van Praag et al. 2002).

[1] Department of Neuroscience, University of New Mexico, School of Medicine, Albuquerque, NM 87131
[2] Department of Chemical Engineering, University of California, Berkeley, CA 94720
[3] The Salk Institute for Biological Studies, 10010 N Torrey Pines Rd, La Jolla, CA 92037

Gage et al.
Stem Cells in the Nervous System:
Functional and Clinical Implications
© Springer-Verlag Berlin Heidelberg 2004

The second question – "What are the functions of these new neurons?" – is more difficult to answer (Barinaga 2003). Since the hippocampus plays an important role in learning and memory and exhibits a high degree of structural plasticity (Squire 1993), neurogenesis in the adult hippocampus may be important for learning and memory in mammals. This hypothesis was first proved by the studies performed on adult songbirds (Goldman and Nottebohm 1983) The levels of neurogenesis in the brains of adult birds are highest in the season in which adult birds need to learn new songs. Later, studies on rodents provided further proof, showing that higher neurogenesis levels are associated with better learning ability in mice of different genetic backgrounds and in mice housed in different environments (Kempermann and Gage 2002; van Praag et al. 1999b). On the other hand, treatment of rats with methylazoxymethanol acetate (MAM), a drug that blocks neurogenesis by selectively killing proliferating cells, resulted in reduced adult neurogenesis and also reduced the learning ability of rats (Shors et al. 2002). In addition, mice lacking MBD1 have reduced neurogenesis and reduced spatial learning (Zhao et al., submitted for publication). Stress causes decreased hippocampal cell proliferation, resulting in learning impairments (Gould et al. 1991; Lemaire et al. 2000). So far the link between neurogenesis and hippocampal functions such as learning remains correlative rather than causal. More detailed research and better technology will be necessary to clearly demonstrate the functions of adult neurogenesis.

Factors affecting in vivo neurogenesis

Much research effort has been invested into understanding the mechanism and regulation of adult neurogenesis, and considerable progress has been made over the past 10 years (Gage 2002). Adult neurogenesis is a complex phenomenon and is affected by many factors. Some of these factors are discussed here and are outlined in Table 1.

Genetic background

It has been shown that mice of different genetic backgrounds have different levels of neurogenesis. For example, mice of varying genetic backgrounds display different levels of cell proliferation and cell survival. When comparing C57B/L6, BALB/c, CD1 and 129/SvJ strains of mice, C57BL6 mice have the highest proliferation rate, whereas CD1 mice have the highest cell survival rate. 120/SvJ mice have the lowest proliferation and survival rates and also the lowest neuronal differentiation and the most gliogenesis of all these mice (Kempermann et al. 1997a). Furthermore, among A/J, C3H/HeJ, and DBA/2J mice, A/J mice have a significantly higher proliferation rate, whereas 3H/HeJ have the highest survival rate. DBA/2J mice also have significantly lower neurogenesis and higher

Table 1. Factors affecting *in vivo* cell proliferation and neurogenesis in adult hippocampus and SVZ/olfactory bulb*.

Factor	Proliferation	Glial genesis	Neurogenesis	References
Genetic background	yes	yes	yes	(Kempermann and Gage, 2002; Kempermann et al., 1997a)
FGF-2	no change increase*	no change increase*	no change increase*	(Kuhn et al., 1997; Wagner et al., 1999)
EGF	no change	increase	decrease	(Craig et al., 1996; Kuhn et al., 1997)
IGF	increase	no change	increase	(Aberg, 2000)
VEGF	increase	n.d.	increase	(Jin et al., 2002)
BDNF	increase*	n.d.	increase*	(Lee et al., 2002)
Serotonin	increase	n.d.	–	(Jacobs et al., 2000)
Norepinephrine (DSP-4)	increase	no change	no change	(Kulkarni et al., 2002)
Glutamate	decrease	n.d.	n.d.	(Cameron et al., 1995; Gould, 1994)
(antagonist: MK801)	increase	n.d.	increase	(Cameron et al., 1998)
Serotonin	increase	n.d.	increase	(Gould, 1999; Jacobs et al., 2000)
Stress	decrease	n.d.	n.d.	(Gould et al., 1997)
Glucocorticoids	decrease	n.d.	n.d.	(Cameron and Gould, 1994)
Adrenalectomy	increase	n.d.	increase	(Cameron and McKay, 1999)
Estrogen	increase	no change	no change	(Tanapat et al., 1999)
Prolactin	increase*	n.d.	increase*	(Shingo et al., 2003)
CAMP/CREB	increase	n.d.	no change	Nakagawa JN 2002
Methamphetamine	decrease	n.d.	n.d.	(Teuchert-Noodt et al., 2000)
Opiate/heroin/ morphine	decrease	n.d.	n.d.	(Eisch et al., 2000)
Enriched environment	no change	no change	increase	(Kempermann et al., 1997b)
Wheel running	increase	no change	increase	(van Praag et al., 1999)
Learning	no change	n.d.	increase no change	(Gould et al., 1999; van Praag et al., 1999)
Dietary restriction	increase	n.d.	n.d.	(Lee et al., 2002)

Table 1. *Continued*

Factor	Proliferation	Glial genesis	Neurogenesis	References
Aging	decrease	n.d.	decrease	(Kuhn et al., 1996)
Vitamin E deficiency	increase	n.d.	n.d.	(Ciaroni et al., 1999)
Traumatic brain injury	increase	increase	increase	(Kernie et al., 2001)
Epilepsy	increase	n.d.	increase	(Parent et al., 1997)
Stroke/ischemia	increase	n.d.	increase	(Liu et al., 1998; Nakatomi et al., 2002)
γ-irradiation	decrease	increase	decrease	(Monje et al., 2002)
Lesion (cortex)#	increase#	increase#	increase#	(Magavi et al., 2000)

Note: *, In SVZ/olfactory bulb only; #, In cortex only; n.d., not determined

astrocyte genesis than other strains (Kempermann and Gage 2002). A second major difference demonstrated in different strains of mice is their response to environmental stimulation. When housed in an enriched environment, C57BL6 mice have increased cell survival without a change in cell proliferation, whereas 129Sv/J have increases in both proliferation and survival (Kempermann et al. 1998a, 1997a).

Growth factors and trophic factors

Several growth factors and trophic factors have been shown to affect adult neurogenesis, including basic fibroblast growth factor (FGF-2), epidermal growth factor (EGF), brain-derived neurotrophic factor (BDNF), and insulin-like growth factor I (IGF-I). Two of the major growth factors that influence NPC proliferation both in vivo and in vitro are FGF-2 and EGF. Intraventricular infusion of FGF-2 and EGF results in increased neurogenesis in the SVZ but not in the dentate gyrus (DG) of the hippocampus (Craig et al. 1996; Kuhn et al. 1997; Wagner et al. 1999). FGF-2 has been shown to be necessary for increased neurogenesis after ischemic insult in the hippocampus, because in the FGF-2 knockout mice, such increased neurogenesis is markedly reduced (Yoshimura et al. 2001). Physical exercise results in increased neurogenesis and increased expression of BDNF, FGF-2 and IGF-1 (Russo-Neustadt et al. 2000; Carro et al. 2000; Gomez-Pinilla et al. 1997). The physical exercise-induced increase in hippocampal neurogenesis is mediated by the increased uptake of IGF-1 into brain from serum (Trejo 2001). Peripheral infusion of IGF-1 can increase adult neurogenesis (Aberget al. 2000), and IGF-1 can also reverse the aging-related reduction in neurogenesis (Lichtenwalner et al. 2001). Intraventricular administration of BDNF increases the number of new neurons in the adult olfactory bulb (Zigova et al. 1998). Dietary restriction results in increased cell proliferation and survival in DG and an increased expression level of BDNF and NT-3 (Lee et al. 2002). Vascular endothelial growth factor (VEGF) can increase proliferation of NPCs both in vitro and in vivo. Intraventricular administration of VEGF causes increased cell proliferation in both SVZ and DG (Jin et al. 2002). Increased neurogenesis in adult songbirds is also associated with increased levels of BDNF, VEGF and VEGF receptor (R2; Louissaint et al. 2002). Dietary restriction, which increases the expression of BDNF and NT-3, can increase cell proliferation in DG (Lee et al. 2002).

Neurotransmitters

Neurotransmitters also play important roles in regulating adult neurogenesis. Glutamate and its agonist have been shown to inhibit cell proliferation in DG, and its antagonist has just the opposite effect (Cameron et al. 1995, 1998b; Gould 1994). Prolonged exposure to monoamine activators can enhance the level of cell proliferation in rats. Serotonin has been shown to stimulate granule cell production (Gould 1999). Serotonin 5-HT (1A) receptor antagonist administration results in decreased cell proliferation in the DG (Radley and Jacobs 2002). The antidepressant effect of serotonin may occur through its effect in augmenting hippocampal neurogenesis (Jacobs et al. 2000). Deletion of norepinephrine by neurotoxin DSP-4 results in a 63% reduction in DG cell proliferation, with no influence on the percentage of neuronal differentiation from proliferating cells (Brezun and Daszuta 1999; Kulkarni et al. 2002). Nitric oxide can promote cell proliferation and migration in both SVZ and DG (Zhang et al. 2001).

Hormonal factors

Stress and its concomitant increase in glucocorticoid levels can reduce the level of neurogenesis in adult rodents (Gould 1992; reviewed by McEwen 1999). Adult neurogenesis in SVZ (Tropepe V 1997 and DG (Seki and Arai, 1995; Kuhn 1996) decreases with age in rodents. This aging-associated decrease can be reversed by adrenalectomy (Cameron and McKay 1999). Testosterone has been shown to increase neurogenesis in birds (Louissaint et al. 2002). Estrogen causes a transient increase in cell proliferation levels in rats (Tanapat et al. 1999), and this process may be mediated by the 5-HT pathway (Banasr et al. 2001). Increased prolactin levels during pregnancy have been associated with increased neurogenesis in SVZ of mice (Shingo et al. 2003). Thyroid hormone has an effect on in vitro differentiation of adult NPCs (Palmer et al. 1995). Increased second messenger cAMP and phosphorylation of its downstream effector, CREB, have been shown to increase neurogenesis (Nakagawa et al. 2002). The effect of serotonin could be mediated through cAMP second messenger cascade (Mendez et al. 1999).

Environment

Enriched environment and physical exercise can increase neurogenesis and spatial learning ability in mice (Kempermann et al. 1997a; van Praag et al. 1999a). Enrichment can partially reverse the reduced neurogenesis level in aged mice (Kempermann et al. 1998b). Enrichment can also provide a neural protective effect against seizure (Auvergne et al. 2002). Even though both enriched environment and physical exercise (running) have been shown to increase the level of neurogenesis, the mechanisms underlying these changes seem to be different. Enrichment increases new cell survival in adult DG but does not affect cell proliferation level, whereas physical exercise increases both, with the same net effect on neurogenesis (Kempermann et al. 1997a; van Praag et al. 1999a).

Diseases/injury conditions and drug abuse

Neurogenesis levels change during the course of CNS diseases and following insults. Ischemia increases cell proliferation in both SVZ and DG, and such an increase is FGF-2-dependent (Liu et al. 1998; Nakatomi et al. 2002; Yoshimura et al. 2001). Seizures have been shown to increase neurogenesis (Parent et al. 1997). Furthermore, CNS injuries can induce neurogenesis in normally non-neurogenic regions (Magavi et al. 2000). Traumatic brain injury induces astrocytogenesis at the proximal site and increased neurogenesis at the distal site in DG (Kernie et al. 2001). Irradiation causes decreased NPC proliferation and neuronal differentiation in rat hippocampus (Monje et al. 2002). Methamphetamine transiently decreases cell proliferation in the DG of the adult gerbil (Teuchert-Noodt et al. 2000). Chronic administration of morphine or heroin can reduce neurogenesis in the adult hippocampus (Eisch et al. 2000), a finding that may explain the poor memory of people with long-term drug abuse (Guerra 1987).

Functions of developmental genes in adult neurogenesis

It is increasingly apparent that genes that are currently best known for their roles in development acquire new and important roles in adulthood, including the regulation of adult neural stem cell function. Furthermore, in some cases, the overexpression or ablation of such genes elicits major developmental phenotypes that complicate studying their role in adulthood, and new methods to genetically modify adult animals must therefore be employed.

Sonic hedgehog (Shh) was first discovered as a crucial regulator of nervous system and limb development (Jessell and Lumsden 1997). Its activity has subsequently been found to be critical in regulating numerous other processes in the developing organism, particularly in the nervous system where it controls midbrain and ventral forebrain neuronal differentiation (Jessell and Lumsden 1997; Ruiz i Altaba et al. 2002a,b). In addition to its regulation of differentiation, however, Shh has been found to control the proliferation of numerous cell populations, including granule neuron precursor cells in the cerebellum, retinal precursors, cells of the epidermis, and other examples (Wechsler-Reya and Scott 1999; Ruiz i Altaba et al. 2002a,b). In light of these numerous roles, it is not surprising that homozygous null-Shh mutation is embryonic lethal (Chiang et al. 1996), and transgenic mice overexpressing Shh are susceptible to tumor formation early in development (Oro et al. 1997), results that complicate the use of these transgenics to analyze its functions in adulthood.

We have recently found that Shh regulates adult hippocampal neural progenitor proliferation in vitro and in vivo (Fig. 1). The addition of recombinant Shh to adult rat hippocampal progenitors in culture stimulates their proliferation in a dose- and time-dependent fashion (Fig. 1a). Furthermore, the use of an adeno-associated viral vector to overexpress Shh in the adult rat hippocampus tripled the number of proliferating cells, and at a later time point these animals had a 3-fold higher number of newborn neurons (Fig. 1b). Finally, cyclopamine, a pharmacological inhibitor of Shh signaling (Berman et al. 2002) reduced the number of proliferating cells in the hippocampal subgranular zone by a factor of two. This study indicates that Shh regulates adult neurogenesis and represents one of its few known roles beyond development (Lai et al. 2003)

Genes that are important during development can regulate not only progenitor proliferation but also their differentiation. Noggin is a polypeptide that binds to and thereby inhibits the function of bone morphogenetic proteins (BMPs). It was originally identified for its regulation of Xenopus neurulation (Smith and Harland 1992), and it has subsequently been found to be important in the development of the neural tube and somite (McMahon et al. 1998). Alvarez-Bullya and colleagues found that BMPs cell-autonomously repress neuronal and promote glial differentiation of adult progenitors in the SVZ. Furthermore, the overexpression of noggin using an adenoviral vector promoted the neuronal differentiation of progenitors grafted into the striatum (Lim et al. 2000). The combined use of multiple factors to control stem cell proliferation and differentiation, such as Shh and noggin, may be a promising approach towards neural regeneration.

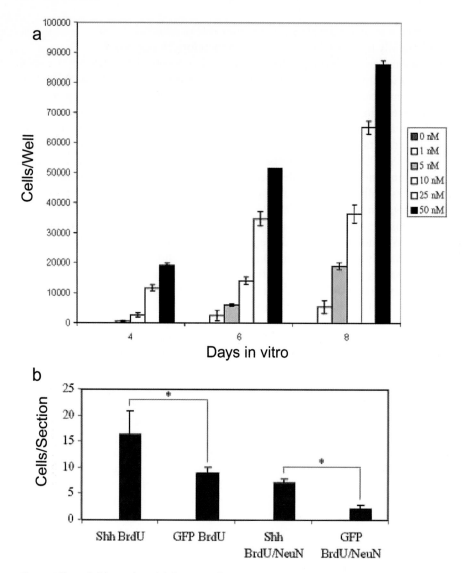

Fig. 1. Effect of Shh on the proliferation of NPCs in vitro and in vivo. (a) Shh induction of NPC proliferation in vitro. Cells were expanded at the concentrations and times indicated, and the final cell number per well is plotted. Each value represents the average of three points compared to the standard curve, with error bars representing standard deviations. (b) Effect of Shh on NPC proliferation and differentiation in vivo. Quantification of the average number of cells in the dentate gyrus and hilar region per section that are BrdU+ and BrdU+/NeuN+ three weeks after the completion of BrdU injections (*P < 0.05). Animals were injected with AAV-Shh or AAV-GFP. (Modified from Lai et al. 2003)

Knowledge obtained from transgenic mice studies

In addition to ectopic overexpression, genetic ablation approaches have also led to the identification of several novel regulators of neurogenesis, as summarized in Table 2. For example, transgenic mutant mice have been used to study the roles of several extracellular signaling molecules. First, the analysis of FGF-2 mutant knockout mice helped confirm the role of this growth factor as a controller of adult hippocampal neurogenesis, as discussed above (Yoshimura et al. 2001). Next, although its function is not fully elucidated, mCD24 is a glycosylphospha tidylinositol-anchored protein widely expressed during development, but only in neurogenic regions of the adult CNS, and it is believed to be involved in cell migration and signaling. mCD24-/- mutant mice exhibited significantly higher rates of progenitor cell proliferation both in the DG and SVZ, indicating that this signaling molecule represses progenitor expansion by an unknown mechanism (Belvindrah et al. 2002).

In addition, although their biological roles are incompletely understood, presenilins are notorious for their involvement in the etiology of Alzheimer's disease. Homozygous mutant presenilin-1 (PS1) animals suffer from severe embryonic abnormalities and embryonic lethality (Shen et al. 1997), so studying its loss-of-function phenotype in adult organisms required the development of an inducible knockout. Tsien and colleagues employed the *Cre-loxP* system, with *Cre* expression driven from the α-calcium-calmodulin-dependent kinase II (CaMKII) promoter, to delete PS1 in the postmitotic neurons in the

Table 2. Studies involving transgenic animals to study cell proliferation and neurogenesis in adult hippocampus and SVZ.

Gene	Transgenic Type	Phenotype	Reference
FGF-2	Homozygous null	Reduced DG proliferation upon injury	(Yoshimura et al., 2001)
mCD24	Homozygous null	Increased DG and SVZ proliferation	(Belvindrah et al., 2002)
PS-1	Conditional null	Reduced DG proliferation in enriched environment	(Feng et al., 2001)
PS-1	Overexpression	Reduced DG proliferation	(Wen et al., 2002)
CREB	Conditional dom. neg.	Reduced DG proliferation	(Nakagawa et al., 2002)
E2F1	Homozygous null	Reduced DG and SVZ proliferation	(Cooper-Kuhn et al., 2002)
P27Kip1	Homozygous null	Increased SVZ proliferation	(Doetsch et al., 2002)
CCg	Homozygous null	Reduced DG proliferation	(Taupin et al., 2000)
MBD1	Homozygous null	Decreased neurogenesis	(Zhao, 2003)

forebrain. Under basal conditions, no changes in the rates of neurogenesis were observed in these animals; however, they had somewhat reduced levels of hippocampal neurogenesis induction by environmental enrichment (Feng et al. 2001). This study indicates that PS1 may promote adult neurogenesis in enriched environments, though a subsequent study showing that transgenics overexpressing PS1 also had reduced numbers of hippocampal progenitors implies that PS1 regulation of neurogenesis is more complex (Wen et al. 2002).

In addition to extracellular signaling molecules, intracellular signal transducers and cell cycle regulators have also been implicated in controlling adult neural stem cell proliferation. CREB is well recognized for its role in learning and memory and neuronal survival signal transduction (Silva et al. 1998). Nakagawa et al. (2002) placed the tetracycline-regulated transactivator under the control of the CamKII promoter in order to inducibly express a dominant negative CREB in the forebrain, which resulted in a 35% reduction in the number of proliferating cells in the DG. Another transcription factor, E2F1, also appears to be responsible for implementing proliferative signals in progenitor cells. The E2F family of transcription factors acts downstream of mitogenic signals (Zhu et al. 2003) and controls the expression of key cell cycle regulators, including cyclins and enzymes involved in nucleotide biosynthesis and DNA replication (Trimarchi and Lees 2002). Homozygous null-E2F1 mutants have significantly reduced progenitor cell proliferation in both the DG and SVZ (Cooper-Kuhn et al. 2002). Finally, ablation of p27Kip1, an inhibitor of cyclin-dependent kinase 2, results in faster proliferation of transit-amplifying cells in the SVZ. This increased proliferation is accompanied by higher levels of apoptosis and also appears to occur at the expense of lineage progression to neuroblasts (Doetsch et al. 2002).

Genetic approaches have therefore led to significant advances in the identification and investigation of a number of modulators of adult neurogenesis. Transgenic and mutant knockout animals permit analysis in cases where a strong developmental phenotype does not preclude the study of adult function. Furthermore, gene delivery, inducible expression systems, and conditional knockouts using the tissue-specific expression (Tsien et al. 1996) or efficient viral vector delivery of *Cre* (Kaspar et al. 2002) are proving to be increasingly powerful tools for genetic studies at developmental times and tissue locations of choice.

Epigenetic factors

DNA methylation at CpG dinucleotides is one of the major epigenetic modifications in the mammalian genome. DNA methylation is involved in transcriptional repression of inactive X-chromosome, imprinted genes, and endogenous retrovirus (see review by Bird 2002). DNA methylation can also regulate the expression of specific genes, such as the expression of GFAP gene in astrocytes (Takizawa et al. 2001). DNA methylation regulates gene expression through two mechanisms. Methylation at CpG sites blocks the binding of transcription factors and leads to transcriptional inactivation. In the second mechanism, methyl-CpGs are bound by a family of methyl-CpG binding proteins (MBDs), including MBD1, 2, 3, 4, and MeCP2. Binding of MBDs and further recruitment of histone deacetylase (HDAC) repressor complexes result in histone deacetylation and inactive chromatin structures that are repressive for transcription (Bird 2002). The most extensively studied member of this family is MeCP2, the mutation of which causes neurological deficits in both humans (Rett Syndrome; Amir et al. 1999) and rodents (Chen et al. 2001; Guy et al. 2001; Shahbazian 2002). Despite extensive efforts to understand the etiology of epigenetic-related diseases (Rett Syndrome and ICF syndrome), the mechanism by which such mutations cause neurological deficits is still not known.

By studying mice lacking MBD1, a member of the MBD family, we established a link between DNA methylation, genomic stability and hippocampal neurogenesis (Fig. 2). Extensive in vitro studies have shown that MBD1 binds specifically to methylated gene promoters through its MBD domain and carries out transcriptional repression that requires its trans-repression domain (TRD; Fujita et al. 2000; Ohki et al. 2001). This process requires an unknown HDAC that is different from the HDAC1 mediating the MeCP2 functions (Ng et al. 2000). The MBD1 is expressed throughout brain regions, with high density in adult hippocampus (Fig. 2a; Zhao et al. 2003). At the cellular level, MBD1 is expressed in neurons but not in glia in the adult mouse brain. In the DG of the hippocampus, the highest expression of MBD1 is in a group of NeuN-negative cells, some of which are also co-labeled with an immature cell marker, nestin (Fig. 2b,c). We have found that MBD1$^{-/-}$ mice have reduced neurogenesis and reduced spatial learning. MBD1$^{-/-}$ mice had no detectable developmental defects and appeared healthy throughout life. However, we found that MBD1$^{-/-}$ neural stem cells exhibited reduced neuronal differentiation and increased genomic instability. Furthermore, MBD1$^{-/-}$ mice had decreased neurogenesis, impaired spatial learning and a significant reduction in long-term potentiation in the DG of the hippocampus (Fig. 2d, e; Zhao et al. submitted for publication). Our work suggests that DNA methylation is important for maintaining genomic stability and normal hippocampal neurogenesis (Fig. 2f).

Neural stem cells in vitro

The identification and isolation of multipotent NPCs from the adult CNS have provided an explanation of the cellular basis for adult neurogenesis. To date, NPCs can be isolated from many different brain regions (Lie et al. 2002; Palmer et al. 1999; Shihabuddin et al. 2000; Weiss et al. 1996). Clonal analysis indicates that these cells are multipotent in in vitro culture systems (Shihabuddin et al. 2000). However, when grafted into the adult brain, these cells differentiate into

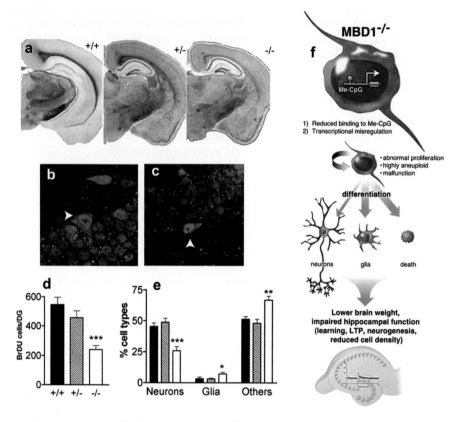

Fig. 2. Mice lacking MBD1 have reduced adult hippocampal neurogenesis. (**a**) X-gal staining showing MBD1 expression pattern in MBD1$^{+/-}$ (+/-) and MBD1$^{-/-}$ mice (-/-) brain. X-gal staining is negative in WT mice brain (+/+). (**b**) MBD1 protein, detected by β-gal expression, is localized in NeuN-positive granule cells of DG (arrowhead) but not in GFAP-positive astrocytes. Some NeuN-negative cells in the hilar region (asterisk) and in the subgranular zone (arrow) express MBD1 at higher levels than NeuN-positive cells in the granule cell layers. (**c**) MBD1 is expressed in some of the nestin-positive cells in the subgranular zone of the DG (arrowhead). (Scale bars = 20 μm). (**d**) Survival of newborn cells in DG is reduced in MBD1$^{-/-}$ mice (p<0.001). (**e**) Phenotype summary of BrdU-positive cells. There is a significant reduction in the percentage of new neurons (p=0.001) and an increase in the percentage of new glia (p<0.05) and unknown cell types (P<0.01) in MBD1$^{-/-}$ mice. Solid bars: WT mice; etched bars, MBD1$^{+/-}$ mice; white bars: MBD1$^{-/-}$ mice. Asterisks: statistically significant: *, p<0.05; **, P<0.01; ***, P<0.001. (**f**) Hypothetical model demonstrating how lacking MBD1 can affect neural functions. (Modified from Zhao et al., submitted for publication)

glial cells in most brain regions, except in the DG and SVZ, where they can differentiate into neurons (Table 3; Lie et al. 2002; Shihabuddin et al. 2000).

FGF-2 and EGF are two potent mitogens for NPCs. During early development, NPCs are responsive only to FGF-2, not EGF. At later stages of development, EGF-responsive NPCs appear (Represa et al. 2001; Temple 2001); EGF-responsive

Table 3 Phenotypic distribution of cloned spinal cord progenitor cell transplants

Phenotype	Spinal Cord (%)	Hippocampus	
		GCL (%)	MCL+CA1+CA3 (%)
NeuN	0	48	3
Calb	ND	44	0
NG2	14	22	26
Rip	3	3	3
GFAP	6	3	3

Relative expression of neuronal and glial markers by transplanted adult spinal cord progenitor cells in the spinal cord and hippocampal formation at 6 weeks after transplantation. Sections were triple-labeled with BrDU, neuronal marker [calbindin (Calb) and NeuN], a glial progenitor marker (NG2), and an astrocytic marker (GFAP), and an oligodendrocyte marker (Rip) and analyzed by confocal microscopy. MCL: molecular cell layer; ND, not determined (Shihabuddin et al., 2000).

NPCs can be isolated from SVZ (Morshead et al. 1994). FGF-2 is a potent mitogen for NPCs isolated from adult hippocampus and other brain regions (Palmer et al. 1999). These two growth factors may have effects on two distinct populations of mouse NPCs. EGF and FGF-2 have synergistic proliferation effects on NPCs isolated from adult mouse brain (Ray, unpublished observation).

Both conditioned media and FGF-2 are required for ANCs to survive when cultured at low density. A glycosylated form of Cystatin (CCg) has been shown to be an essential component of the conditioned media. CCg is expressed in the subgranular layer of the DG in adult hippocampus, and exogenous CCg can increase NPC cell proliferation both in vitro and in vivo (Taupin and Gage 2002). Other growth factors, such as NT-3 and BDNF, have been shown to mildly increase neuronal differentiation without changing the proliferation rate of cultured NPCs (Palmer et al. 1995). NGF has been shown to affect NPC in vitro proliferation/differentiation (Cameron et al. 1998a). IGF-1 can increase NPC cell proliferation in vitro and can also influence NPC cell fate in terms of what type of neurotransmitters the differentiated neurons secrete (Anderson et al. 2002). TGF-β family members and its subfamily, BMPs, have been shown to affect the NPC cell fate (Cameron et al. 1998a).

It is of great interest to understand what factors can influence NPC lineage determination. Many studies have been performed using an in vitro cell culture system because it is easier to control for individual factors. However, most differentiation treatments do not result in 100% desired phenotype. Retinoic acid (RA) and a cAMP pathway activator, forskolin, have been used to differentiate NPCs into neuronal lineage; however, only up to 50% of differentiated cells are labeled with neuronal cell marker (Palmer et al. 1997; Gage lab observation).

Both fetal bovine serum alone or in combination with BMP and LIF have been used to differentiate embryonic NPCs into about 40% astrocytes (Nakashima et al. 1999). Despite extensive efforts, we still cannot instruct NPCs to become 100% of the cell types that we want. However, we are making slow and steady progress.

Model 1: Intrinsic properties of NPCs determine the cell fate

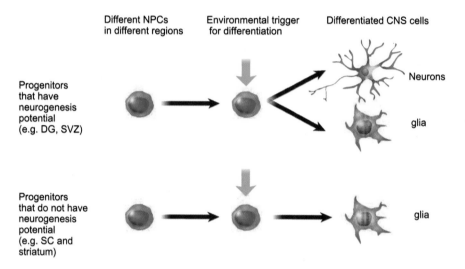

Model 2: Environmental cues determine NPC cell fate

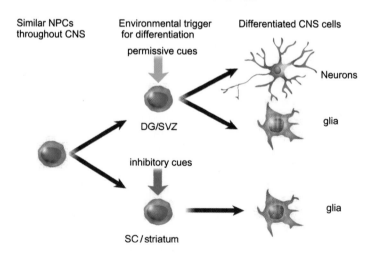

Fig. 3. Two possible mechanisms that control NPC cell fate in vivo.

Why does neurogenesis occur only in certain regions of the adult brain?

A fundamental question in understanding how neurogenesis is regulated is why neurogenesis only exists in restricted areas in the adult mammalian brain. At the molecular level, there are two possible explanations: NPC fate is dictated by 1) differences in the intrinsic properties of different NPCs residing in the neurogenic region vs. non-neurogenic region and 2) different environmental cues elicited by the neurogenic environment vs, the non-neurogenic environment (Fig. 3).

Both mechanisms may play critical roles in the process. NPCs isolated from non-neurogenic regions of the brain had to be cultured in the presence of FGF-2 for several passages before they could be differentiated into neurons, whereas NPCs isolated from DG could be differentiated into neurons immediately after being isolated from the brain (Palmer et al. 1999). Furthermore, NPCs isolated from the hippocampus could be differentiated into neurons, astrocytes and oligodendrocytes when treated with RA and 0.5% FBS for two weeks, resulting in about 10% neurons and 10-20% oligodendrocytes (Palmer et al. 1997; other unpublished data in Gage lab). NPCs isolated from spinal cord treated with the same condition, however, gave rise to over 50% oligodendrocytes, 10% neurons and less than 10% astrocytes (Gage lab observation). These observations indicate that even though multipotent NPCs can be isolated from many different brain regions, there are some intrinsic differences in freshly isolated NPCs. Such differences can be diminished in cell culture in the presence of growth factors.

Stronger experimental evidence supports the hypothesis that environmental factors are critical in determining NPC fate. First, when cultured multipotent NPCs isolated from SC were grafted back into the brain, they differentiated into neurons only in DG and SVZ. In SC and other non-neurogenic brain regions, grafted NPCs only differentiated into glial cells, indicating that the local environments played critical roles in determining NPC cell fate (Table 3; Lie et al. 2002; Shihabuddin et al. 2000). Second, Song et al. (2002a) co-cultured GFP-labeled NPCs with primary astrocytes and found that astrocytes isolated from the hippocampus and SC of newborns and adults have distinct effects on NPC neuronal differentiation in vitro. Astrocytes from the newborn hippocampus have the strongest neuronal lineage-promoting effect, whereas the astrocytes from adult SC have the strongest neurogenesis-inhibiting effects (Fig. 4). Because both conditioned media from primary hippocampal astrocytes and lightly fixed astrocytes have only partially greater neurogenesis-promoting effect than the pure astrocytes, Song et al. (2002a) concluded that both membrane-bound and secreted factors from astrocytes have effects on co-cultured NPCs. Since, during postnatal neurogenesis, NPCs are in close contact with surrounding glial cells, including astrocytes, it is likely that local astrocytes play an important role in determining NPC cell fate in vivo.

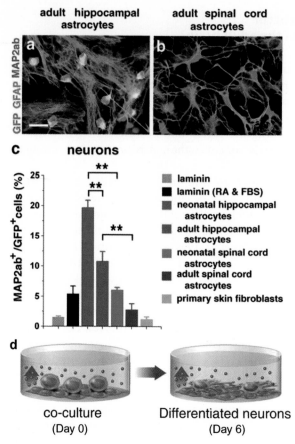

Fig. 4. Mature astrocytes from adult hippocampus, but not adult spinal cord, promote neurogenesis from adult stem cells. (**a, b**) Differentiation of GFP+ adult neural stem cells in co-culture with astrocytes derived from adult hippocampus (**a**) or adult spinal cord (**b**). Cells in six-day cultures were stained for MAP2ab and GFAP. Scale bar, 5 mm.(**c**) Quantification of the percentage of MAP2ab+ and GFP+ neurons in different conditions (six-day culture). Data shown are mean values ∧ s.e.m. from four to eight experiments in parallel cultures. Significant differences between results for astrocytes from hippocampus and spinal cord are indicated by a double asterisk (P< 0.01, t-test). (**d**) Schematic diagram showing experimental set-up. (Modified from Song et al. 2002a)

To search for both intrinsic and extrinsic factors that affect neurogenesis, we applied state-of-the-art gene expression profiling and functional genomics approaches. Two experiments have been performed. One analyzed regional gene expression differences between neurogenic (DG) and non-neurogenic regions (CA1, and SC). We identified sets of genes with sub-regional specific expression patterns, which we further confirmed by using real time quantitative PCR and in situ hybridization. A montage of regional specific genes shows that gene expression patterns correlate with anatomical boundaries (Fig. 5: Zhao

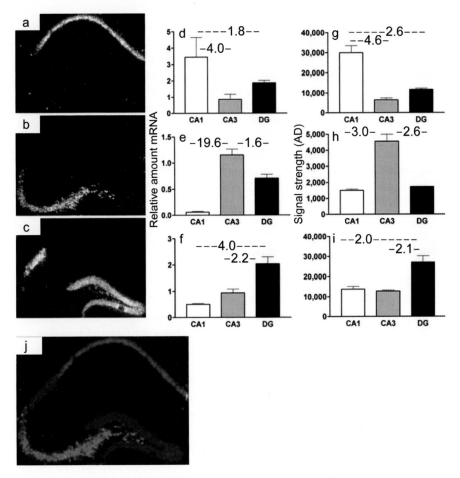

Fig. 5. Confirmation of region-specific genes in the hippocampus by in situ hybridization and Real Time quantitative PCR. (**a,d,g**) NOV; (**b,e,h**) PKC-δ; (**c,f,i**) PCP-4. (**a-c**) in situ hybridization of serial sections; (**d-f**) Real Time quantitative PCR data. The relative amounts of mRNA for each gene were normalized to the amounts of HPRT mRNA (see Materials and Methods for detail); (**g-i**) Signal strength: AD value from microarray analysis, indicating relative expression level. Each bar represents an average of AD values obtained from the two true duplicates. Note that the fold changes obtained from Real Time PCR and microarray analysis are in general agreement. (**j**) False-colored montage of in situ hybridization from serial sections shown in **a, b, c**, demonstrating molecular boundaries between anatomical regions in hippocampus. Red: NOV; Green: PKC-δ; Blue: PCP-4. Note that PCP-4 expression delineated CA2 as a molecularly distinct structure from CA1 and CA3. (Modified from Zhao et al. 2001)

et al. 2001). A search for subregional specific promoter elements is underway, with hopes this research will provide useful tools for studying hippocampal functions. Candidate genes that differentially express between neurogenic and non-neurogenic regions have been identified and their effect on neurogenesis is currently being tested using an in vitro culture system.

A second approach analyzed gene expression differences among astrocytes with distinct effects on neurogenesis. This approach will give us an even more definitive answer because only one cell type is analyzed. Our findings show that many genes are expressed differently among these astrocytes (Zhao and Song, unpublished data). Verification of the effects of these genes on neurogenesis is currently underway. The result of this study will provide further understanding of how adult neurogenesis is regulated.

Conclusions and future directions

Two challenges face us:
1. How do we prove that adult neurogenesis is directly linked to hippocampal function, such as learning and memory? To answer this question, we need a more detailed analysis to trace the fate and function of new hippocampal granule neurons. For example, do new neurons form the right connection with the right targets? On the other hand, the use of a specific method to kill hippocampal NPCs in vivo, such as NPC cell type-specific promoter-directed cell death gene expression, will help answer the question of whether destroying adult neurogenesis will result in impaired learning.
2. How do we identify the factors that can control NPC cell fate choice both in vitro and in vivo? We have made slow but steady progress over the past 10 years. The new genomics and proteomics technologies will help to speed up the process dramatically. However, a rate-limiting step is functional analysis of the candidate genes. More efficient and high throughput methods will be necessary to break the bottleneck.

Our long-term goal is to find the molecular keys for neurogenesis in the adult brain so that we can either graft multipotent NPCs into diseased regions or mobilize local stem cells and instruct them to become the numbers and types of neurons needed for CNS repair. To achieve this goal, we first have to understand the molecular mechanism underlying adult neurogenesis. It is a complex and difficult question, but with more and more effort being put into this research and better technology being developed, the answer is on the horizon.

References

Aberg MA, Aberg ND, Hedbacker H, Oscarsson J, Eriksson PS (2000) Peripheral infusion of IGF–I selectively induces neurogenesis in the adult rat hippocampus. J Neurosci 20: 2896–2903

Altman J, Das GD (1965) Autoradiographic and histological evidence of postnatal neurogenesis in rats. J Comp Neurol 124: 319–335

Amir RE, Van den Veyver IB, Wan M, Tran CQ, Francke U, Zoghbi HY (1999) Rett syndrome is caused by mutations in X–linked MECP2, encoding methyl– CpG–binding protein 2. Nature Genet 23: 185–188

Anderson MF, Aberg MA, Nilsson M, Eriksson PS (2002) Insulin–like growth factor–I and neurogenesis in the adult mammalian brain. Brain Res Dev Brain Res 134: 115–122

Auvergne R, Lere C, El Bahh B, Arthaud S, Lespinet V, Rougier A, Le Gal La Salle G (2002) Delayed kindling epileptogenesis and increased neurogenesis in adult rats housed in an enriched environment.Brain Res 954: 277–285

Banasr M, Hery M, Brezun JM, Daszuta A (2001) Serotonin mediates oestrogen stimulation of cell proliferation in the adult dentate gyrus. Eur J Neurosci 14: 1417–1424

Barinaga, M (2003) Developmental biology. Newborn neurons search for meaning. Science 299: 32–34

Belvindrah R, Rougon G, Chazal G (2002) Increased neurogenesis in adult mCD24–deficient mice. J Neurosci 22: 3594–3607

Berman DM, Karhadkar SS, Hallahan AR, Pritchard JI, Eberhart CG, Watkins DN, Chen JK, Cooper MK, Taipale J, Olson JM, Beachy PA (2002) Medulloblastoma growth inhibition by hedgehog pathway blockade. Science 297: 1559–1561

Bird A (2002) DNA methylation patterns and epigenetic memory. Genes Dev 16: 6–21

Brezun JM, Daszuta A (1999) Depletion in serotonin decreases neurogenesis in the dentate gyrus and the subventricular zone of adult rats. Neuroscience 89: 999–1002

Cameron HA, McKay RD (1999) Restoring production of hippocampal neurons in old age. Nature Neurosci 2: 894–897

Cameron HA, McEwen BS, Gould E (1995) Regulation of adult neurogenesis by excitatory input and NMDA receptor activation in the dentate gyrus. J Neurosci 15: 4687–4692

Cameron HA, Hazel TG, McKay RD (1998a) Regulation of neurogenesis by growth factors and neurotransmitters. J Neurobiol 36: 287–306

Cameron HA, Tanapat P, Gould E (1998b) Adrenal steroids and N–methyl–D–aspartate receptor activation regulate neurogenesis in the dentate gyrus of adult rats through a common pathway. Neuroscience 82: 349–354

Carro E, Nunez A, Busiguina S, Torres–Aleman I (2000) Circulating insulin–like growth factor I mediates effects of exercise on the brain. J Neurosci 20: 2926–2933

Chen RZ, Akbarian S, Tudor M, Jaenisch R (2001) Deficiency of methyl–CpG binding protein–2 in CNS neurons results in a Rett–like phenotype in mice. Nature Genet 27: 327–331

Chiang C, Litingtung Y, Lee E, Young KE, Corden J L, Westphal H, Beachy P A (1996) Cyclopia and defective axial patterning in mice lacking Sonic hedgehog gene function. Nature 383: 407–413

Cooper-Kuhn CM, Vroemen M, Brown J, Ye H, Thompson MA, Winkler J, Kuhn HG (2002) Impaired adult neurogenesis in mice lacking the transcription factor E2F1. Mol Cell Neurosci 21: 312–323

Craig CG, Tropepe V, Morshead CM, Reynolds BA, Weiss S, van der Kooy D (1996) In vivo growth factor expansion of endogenous subependymal neural precursor cell populations in the adult mouse brain. J Neurosci 16: 2649–2658

Doetsch F, Verdugo JM, Caille I, Alvarez–Buylla A, Chao MV, Casaccia–Bonnefil P (2002) Lack of the cell–cycle inhibitor p27Kip1 results in selective increase of transit–amplifying cells for adult neurogenesis. J Neurosci 22: 2255–2264

Eisch AJ, Barrot M, Schad CA, Self DW, Nestler EJ (2000) Opiates inhibit neurogenesis in the adult rat hippocampus. Proc Natl Acad Sci USA 97: 7579–7584

Eriksson PS, Perfilieva E, Bjork–Eriksson T, Alborn AM, Nordborg C, Peterson DA, Gage FH (1998) Neurogenesis in the adult human hippocampus. Nature Med 4: 1313–1317

Feng R, Rampon C, Tang YP, Shrom D, Jin J, Kyin M, Sopher B, Miller MW, Ware CB, Martin GM, Kim SH, Langdon RB, Sisodia SS, Tsien JZ (2001) Deficient neurogenesis in forebrain–specific presenilin–1 knockout mice is associated with reduced clearance of hippocampal memory traces. Neuron 32: 911–926

Fujita N, Shimotake N, Ohki I, Chiba T, Saya H, Shirakawa M, Nakao M (2000) Mechanism of transcriptional regulation by methyl–CpG binding protein MBD1. Mol Cell Biol 20: 5107–5118

Gage FH (2002) Neurogenesis in the adult brain. J Neurosci 22: 612–613

Goldman SA, Nottebohm F (1983) Neuronal production, migration, and differentiation in a vocal control nucleus of the adult female canary brain. Proc Natl Acad Sci USA 80: 2390–2394

Gomez–Pinilla F, Dao L, So V (1997) Physical exercise induces FGF–2 and its mRNA in the hippocampus. Brain Res 764: 1–8

Gould E (1994) The effects of adrenal steroids and excitatory input on neuronal birth and survival. Ann NY Acad Sci 743:73–92; discussion 92–93

Gould E (1999) Serotonin and hippocampal neurogenesis, Neuropsychopharmacology 21: 46S–51S

Gould E, Woolley CS, McEwen BS (1991) Adrenal steroids regulate postnatal development of the rat dentate gyrus: I. Effects of glucocorticoids on cell death. J Comp Neurol 313: 479–485

Gould E, Cameron HA, Daniels DC, Woolley CS, McEwen BS (1992) Adrenal hormones suppress cell division in the adult rat dentate gyrus. J Neurosci 12:3642–3650

Gould E, Reeves AJ, Fallah M, Tanapat P, Gross CG, Fuchs E (1999) Hippocampal neurogenesis in adult Old World primates. Proc Natl Acad Sci USA 96: 5263–5267

Guerra D, Sole A, Cami J, Tobena A (1987) Neuropsychological performance in opiate addicts after rapid detoxification. Drug Alcohol Depend 20: 261–270

Guy J, Hendrich B, Holmes M, Martin JE, Bird A (2001) A mouse Mecp2–null mutation causes neurological symptoms that mimic Rett syndrome. Nature Genet 27: 322–326

Jacobs BL, Praag H, Gage FH (2000) Adult brain neurogenesis and psychiatry: a novel theory of depression. Mol Psychiat 5: 262–269

Jessell TM, Lumsden A (1997) Inductive signals and the assignment of cell fate in the spinal cord and hindbrain In: Cowan WM, Jessell TM, Zipursky SL (eds) Molecular and cellular approaches to neural development. Oxford, Oxford University Press pp 290–333

Jin K, Zhu Y, Sun Y, Mao XO, Xie L, Greenberg DA (2002) Vascular endothelial growth factor (VEGF) stimulates neurogenesis in vitro and in vivo. Proc Natl Acad Sci USA 99: 11946–11950

Kaplan MS, Hinds JW (1977) Neurogenesis in the adult rat: electron microscopic analysis of light radioautographs. Science 197: 1092–1094

Kaspar BK, Vissel B, Bengoechea T, Crone S, Randolph–Moore L, Muller R, Brandon EP, Schaffer D, Verma IM, Lee KF, Heinemann SF, Gage FH (2002) Adeno–associated virus effectively mediates conditional gene modification in the brain. Proc Natl Acad Sci USA 99: 2320–2325

Kempermann G, Gage FH (2002) Genetic influence on phenotypic differentiation in adult hippocampal neurogenesis. Brain Res Dev Brain Res 134: 1–12

Kempermann G, Kuhn HG, Gage FH (1997a) Genetic influence on neurogenesis in the dentate gyrus of adult mice. Proc Natl Acad Sci USA 94: 10409–10414

Kempermann G, Kuhn HG, Gage FH (1997b) More hippocampal neurons in adult mice living in an enriched environment. Nature 386: 493–495

Kempermann G, Brandon EP, Gage FH (1998a) Environmental stimulation of 129/SvJ mice causes increased cell proliferation and neurogenesis in the adult dentate gyrus. Curr Biol 8: 939–942

Kempermann G, Kuhn HG, Gage FH (1998b) Experience–induced neurogenesis in the senescent dentate gyrus. J Neurosci 18: 3206–3212

Kernie SG, Erwin TM, Parada LF (2001) Brain remodeling due to neuronal and astrocytic proliferation after controlled cortical injury in mice. J Neurosci Res 66: 317–326

Kornack DR, Rakic P (1999) Continuation of neurogenesis in the hippocampus of the adult macaque monkey. Proc Natl Acad Sci USA 96: 5768–5773

Kuhn GG, Dickinson-Anson H, Gage FH (1996) Neurogenesis in the dentate gyrus of the adult rat: age–related decrease of neuronal progenitor proliferation. J Neurosci 16: 2027–2023

Kuhn HG, Winkler J, Kempermann G, Thal LJ, Gage FH (1997) Epidermal growth factor and fibroblast growth factor–2 have different effects on neural progenitors in the adult rat brain. J Neurosci 17: 5820–5829

Kulkarni VA, Jha S, Vaidya VA (2002) Depletion of norepinephrine decreases the proliferation, but does not influence the survival and differentiation, of granule cell progenitors in the adult rat hippocampus. Eur J Neurosci 16: 2008–2012

Lai K, Kaspar BK, Gage FH, Schaffer DV (2003) Sonic hedgehog regulates adult neural progenitor proliferation in vitro and in vivo. Nature Neurosci 6: 21–27

Lee J, Seroogy KB, Mattson MP (2002) Dietary restriction enhances neurotrophin expression and neurogenesis in the hippocampus of adult mice. J Neurochem 80: 539–547

Lemaire V, Koehl M, Le Moal M, Abrous DN (2000) Prenatal stress produces learning deficits associated with an inhibition of neurogenesis in the hippocampus. Proc Natl Acad Sci USA 97: 11032–11037

Lichtenwalner RJ, Forbes ME, Bennett SA, Lynch CD, Sonntag WE, Riddle DR (2001) Intracerebroventricular infusion of insulin–like growth factor–I ameliorates the age–related decline in hippocampal neurogenesis. Neuroscience 107: 603–613

Lie DC, Dziewczapolski G, Willhoite AR, Kaspar BK, Shults CW, Gage FH (2002) The adult substantia nigra contains progenitor cells with neurogenic potential. J Neurosci 22: 6639–6649

Lim DA, Tramontin AD, Trevejo JM, Herrera DG, Garcia–Verdugo JM, Alvarez–Buylla A (2000) Noggin antagonizes BMP signaling to create a niche for adult neurogenesis Neuron 28: 713–726

Liu J, Solway K, Messing RO, Sharp FR (1998) Increased neurogenesis in the dentate gyrus after transient global ischemia in gerbils. J Neurosci 18: 7768–7778

Louissaint A, Jr, Rao S, Leventhal C, Goldman SA (2002) Coordinated interaction of neurogenesis and angiogenesis in the adult songbird brain. Neuron 34: 945–960

Magavi SS, Leavitt BR, Macklis JD (2000) Induction of neurogenesis in the neocortex of adult mice. Nature 405: 951–955

McEwen BS (1999) Stress and hippocampal plasticity. Annu Rev Neurosci 22:105–122

McMahon JA, Takada S, Zimmerman LB, Fan CM, Harland R M, McMahon AP (1998) Noggin–mediated antagonism of BMP signaling is required for growth and patterning of the neural tube and somite Genes Dev 12: 1438–1452

Mendez J, Kadia TM, Somayazula RK, El–Badawi KI, Cowen DS (1999) Differential coupling of serotonin 5–HT1A and 5–HT1B receptors to activation of ERK2 and inhibition of adenylyl cyclase in transfected CHO cells. J Neurochem 73: 162–168

Monje ML, Mizumatsu S, Fike JR, Palmer TD (2002) Irradiation induces neural precursor–cell dysfunction. Nature Med 8: 955–962

Morshead CM, Reynolds BA, Craig CG, McBurney MW, Staines WA, Morassutti D, Weiss S, van der Kooy D (1994) Neural stem cells in the adult mammalian forebrain: a relatively quiescent subpopulation of subependymal cells Neuron 13: 1071–1082

Nakagawa S, Kim JE, Lee R, Malberg JE, Chen J, Steffen C, Zhang YJ, Nestler EJ, Duman RS (2002) Regulation of neurogenesis in adult mouse hippocampus by cAMP and the cAMP response element–binding protein. J Neurosci 22: 3673–3682

Nakashima K, Yanagisawa M, Arakawa H, Kimura N, Hisatsune T, Kawabata M, Miyazono K, Taga T (1999) Synergistic signaling in fetal brain by STAT3–Smad1 complex bridged by p300. Science 284: 479–482

Nakatomi H, Kuriu T, Okabe S, Yamamoto S, Hatano O, Kawahara N, Tamura A, Kirino T, Nakafuku M (2002) Regeneration of hippocampal pyramidal neurons after ischemic brain injury by recruitment of endogenous neural progenitors. Cell 110: 429–441

Ng HH, Jeppesen P, Bird A (2000) Active repression of methylated genes by the chromosomal protein MBD1. Mol Cell Biol 20: 1394–1406

Ohki I, Shimotake N, Fujita N, Jee J, Ikegami T, Nakao M, Shirakawa M (2001) Solution structure of the methyl–CpG binding domain of human MBD1 in complex with methylated DNA. Cell 105: 487–497

Oro AE, Higgins KM, Hu Z, Bonifas JM, Epstein EH Jr., Scott MP (1997) Basal cell carcinomas in mice overexpressing sonic hedgehog. Science 276: 817–821

Palmer TD, Ray J, Gage FH (1995) FGF–2–responsive neuronal progenitors reside in proliferative and quiescent regions of the adult rodent brain. Mol Cell Neurosci 6: 474–486

Palmer TD, Takahashi J, Gage FH (1997) The adult rat hippocampus contains primordial neural stem cells Mol Cell Neurosci 8: 389–404

Palmer TD, Markakis EA, Willhoite AR, Safar F, Gage FH (1999) Fibroblast growth factor–2 activates a latent neurogenic program in neural stem cells from diverse regions of the adult CNS. J Neurosci 19: 8487–8497

Parent JM, Yu TW, Leibowitz RT, Geschwind DH, Sloviter RS, Lowenstein DH (1997) Dentate granule cell neurogenesis is increased by seizures and contributes to aberrant network reorganization in the adult rat hippocampus. J Neurosci 17: 3727–3738

Radley JJ, Jacobs BL (2002) 5–HT1A receptor antagonist administration decreases cell proliferation in the dentate gyrus. Brain Res 955: 264–267

Represa A, Shimazaki T, Simmonds M, Weiss S (2001) EGF–responsive neural stem cells are a transient population in the developing mouse spinal cord. Eur J Neurosci 14: 452–462

Ruiz i Altaba A, Palma V, Dahmane N (2002a) Hedgehog–Gli signalling and the growth of the brain. Nature Rev Neurosci 3: 24–33

Ruiz i Altaba A, Sanchez P, Dahmane N (2002b) Gli and hedgehog in cancer: tumours, embryos and stem cells. Nature Rev Cancer 2: 361–372

Russo–Neustadt AA, Beard RC, Huang YM, Cotman CW (2000) Physical activity and antidepressant treatment potentiate the expression of specific brain–derived neurotrophic factor transcripts in the rat hippocampus. Neuroscience 101: 305–312

Seki T, Arai Y (1995) Age–related production of new granule cells in the adult dentate gyrus. Neuroreport 6:2479–2482

Shahbazian MD, Young JI, Yuva–Paylor LA, Spencer CM, Antalffy BA, Noebels JL, Armstrong DL, Paylor R, Zoghbi HY (2002) Mice with truncated MeCP2 recapitulate many Rett Syndrome features and display hyperacetylation of histone H3. Neuron 35: 243–254

Shen J, Bronson RT, Chen DF, Xia W, Selkoe DJ, Tonegawa S (1997) Skeletal and CNS defects in Presenilin–1–deficient mice. Cell 89: 629–639

Shihabuddin LS, Horner PJ, Ray J, Gage FH (2000) Adult spinal cord stem cells generate neurons after transplantation in the adult dentate gyrus. J Neurosci 20: 8727–8735

Shingo T, Gregg C, Enwere E, Fujikawa H, Hassam R, Geary C, Cross JC, Weiss S (2003) Pregnancy–stimulated neurogenesis in the adult female forebrain mediated by prolactin. Science 299: 117–120

Shors TJ, Townsend DA, Zhao M, Kozorovitskiy Y, Gould E (2002) Neurogenesis may relate to some but not all types of hippocampal- dependent learning. Hippocampus 12: 578–584

Silva AJ, Kogan JH, Frankland PW, Kida S (1998) CREB and memory. Annu Rev Neurosci 21: 127–148

Smith WC, Harland RM (1992) Expression cloning of noggin, a new dorsalizing factor localized to the Spemann organizer in Xenopus embryos. Cell 70: 829–840

Song H, Stevens CF, Gage FH (2002a) Astroglia induce neurogenesis from adult neural stem cells. Nature 417: 39–44

Song HJ, Stevens CF, Gage FH (2002b) Neural stem cells from adult hippocampus develop essential properties of functional CNS neurons. Nature Neurosci 5: 438–445

Squire LR (1993) The hippocampus and spatial memory. Trends Neurosci 16: 56–57

Takizawa T, Nakashima K, Namihira M, Ochiai W, Uemura A, Yanagisawa M, Fujita N, Nakao M, Taga T (2001) DNA methylation is a critical cell–intrinsic determinant of astrocyte differentiation in the fetal brain. Dev Cell 1: 749–758

Tanapat P, Hastings NB, Reeves AJ, Gould E (1999) Estrogen stimulates a transient increase in the number of new neurons in the dentate gyrus of the adult female rat. J Neurosci 19: 5792–5801

Taupin P, Gage FH (2002) Adult neurogenesis and neural stem cells of the central nervous system in mammals. J Neurosci Res 69: 745–749

Temple S (2001) The development of neural stem cells. Nature 414: 112–117

Teuchert-Noodt G, Dawirs RR, Hildebrandt K (2000) Adult treatment with methamphetamine transiently decreases dentate granule cell proliferation in the gerbil hippocampus J Neural Transm 107: 133–143

Trejo JL, Carro E, Torres-Alemán I (2001) Circulating insulin–like growth factor I mediates exercise–induced increases in the number of new neurons in the adult hippocampus. J Neurosci 21: 1628–1634

Trimarchi JM, Lees JA (2002) Sibling rivalry in the E2F family. Nature Rev Mol Cell Biol 3: 11–20

Tropepe V, Craig CG, Morshead CM, van der Kooy D (1997) Transforming growth factor–alpha null and senescent mice show decreased neural progenitor cell proliferation in the forebrain subependyma. J Neurosci 17: 7850–7859

Tsien JZ, Chen DF, Gerber D, Tom C, Mercer EH, Anderson DJ, Mayford M, Kandel ER, Tonegawa S (1996) Subregion– and cell type–restricted gene knockout in mouse brain. Cell 87: 1317–1326

van Praag H, Christie BR, Sejnowski TJ, Gage FH (1999a) Running enhances neurogenesis, learning, and long–term potentiation in mice. Proc Natl Acad Sci USA 96: 13427–13431

van Praag H, Kempermann G, Gage FH (1999b) Running increases cell proliferation and neurogenesis in the adult mouse dentate gyrus. Nature Neurosci 2: 266–270

van Praag H, Schinder AF, Christie BR, Toni N, Palmer TD, Gage FH (2002) Functional neurogenesis in the adult hippocampus. Nature 415: 1030–1034

Wagner JP, Black IB, DiCicco–Bloom E (1999) Stimulation of neonatal and adult brain neurogenesis by subcutaneous injection of basic fibroblast growth factor. J Neurosci 19: 6006–6016

Wechsler–Reya RJ, Scott MP (1999) Control of neuronal precursor proliferation in the cerebellum by Sonic Hedgehog Neuron 22: 103–114

Weiss S, Dunne C, Hewson J, Wohl C, Wheatley M, Peterson AC, Reynolds BA (1996) Multipotent CNS stem cells are present in the adult mammalian spinal cord and ventricular neuroaxis. J Neurosci 16:7599–7609

Wen PH, Shao X, Shao Z, Hof PR, Wisniewski T, Kelley K, Friedrich VL Jr, Ho L, Pasinetti GM, Shioi J, Robakis NK, Elder GA (2002)
Overexpression of wild type but not an FAD mutant presenilin–1 promotes neurogenesis in the hippocampus of adult mice. Neurobiol Dis 10: 8–19

Yoshimura S, Takagi Y, Harada J, Teramoto T, Thomas SS, Waeber C, Bakowska JC, Breakefield XO, Moskowitz MA (2001) FGF–2 regulation of neurogenesis in adult hippocampus after brain injury. Proc Natl Acad Sci USA 98: 5874–5879

Zhang R, Zhang L, Zhang Z, Wang Y, Lu M, Lapointe M, Chopp M (2001) A nitric oxide donor induces neurogenesis and reduces functional deficits after stroke in rats. Ann Neurol 50: 602–611

Zhao X, Lein ES, He A, Smith SC, Aston C, Gage FH (2001) Transcriptional profiling reveals strict boundaries between hippocampal subregions. J Comp Neurol 441: 187–196

Zhao X, Veba T, Christie BR, Barkho B, McConnell MJ, Nakashima K, Lein ES, Eadie B, Chun J, Lee K, Gage FH (2003) Methyl-Cp G binding protein 1 is important for neurogenesis and genomic stability in adult neural stem cells. Proc. Nath. Aco. Sci. USA 100(11): 6777-82.

Zhu Y, Jin K, Mao XO, Greenberg DA (2003) Vascular endothelial growth factor promotes proliferation of cortical neuron precursors by regulating E2F expression. FASEB J 17: 186–193

Zigova T, Pencea V, Wiegand SJ, Luskin MB (1998) Intraventricular administration of BDNF increases the number of newly generated neurons in the adult olfactory bulb. Mol Cell Neurosci 11: 234–245.

The Logic of Neural Cell Lineage Restriction: Neuropoiesis Revisited

D.J. Anderson[1], L. Lo, M. Zirlinger[2], G. Choi and *Q. Zhou*

Summary

The neurons and glia of the central and peripheral nervous systems (CNS and PNS) are thought to derive from initially multipotent, self-renewing stem cells. These cells have been assumed to undergo a sequence of lineage restrictions, in which they first generate committed neuronal or glial progenitors, which then generate different subtypes of neurons and glia, respectively. Evidence presented here from both the PNS and CNS suggests that, although the fundamental concept of sequential lineage restriction is likely to be correct, the logic of such restrictions is not the one widely assumed to be true. Rather, we suggest that multipotent progenitors of neurons and glia become fate-restricted to generating different subtypes of neurons and glia, before they become committed to neuronal and glial fates. This pattern of lineage restriction events can be explained in terms of the molecular mechanisms that control the neuron vs. glia fate choice and of those that specify neuronal and glial subtype identity. An outstanding question raised by these studies is the relationship of such multipotent but subtype-restricted progenitor cells in vivo to self-renewing CNS stem cells that have been described in vitro.

Introduction

A major problem in understanding neural development is determining the logic and mechanisms by which undifferentiated progenitor cells generate the diverse neuronal and glial subtypes of the nervous system. An assumption shared by many in the field is that the entire nervous system develops from a population of multipotent, self-renewing stem cells, which in the CNS generate neurons, astrocytes and oligodendrocytes. Indeed such cells have been identified and isolated in cultures derived from the fetal brain, either as floating neurospheres or as monolayer cultures (reviewed in Gage 2000; Temple 2001). However, it is

---

[1] Howard Hughes Medical Institute, California Institute of Technology, Pasadena, CA USA

[2] Present address: Department of Biochemistry and Molecular Biology, Harvard University, 16 Divinity Ave., Cambridge, MA 02138

Gage et al.
Stem Cells in the Nervous System:
Functional and Clinical Implications
© Springer-Verlag Berlin Heidelberg 2004

not yet clear whether these stem cells represent the majority, or only a minority, of progenitor cells in the embryonic ventricular zone in vivo. Neither is it yet clear whether a clonogenic population of CNS stem cells can generate all of the different neuronal subtypes in the CNS, or whether stem cells from different regions of the brain exhibit regional restrictions in their developmental capacities (Hitoshi et al. 2002).

Nevertheless, based on the assumption that at least some embryonic neural progenitor cells behave as stem cells in vivo, as well as in vitro, it is important to understand the logic and mechanisms by which these cells generate their differentiated progeny. By "logic" we mean the "decision tree" that governs the production of different cell types. In the hematopoietic system, for example, stem cells generate a series of developmentally restricted intermediate progenitors, beginning with the segregation of common lymphoid (Scherer et al. 2000) and myeloid progenitors (Scherer et al. 2000; Fig. 1A). These progenitors then undergo further lineage restriction events until they generate their differentiated derivatives. In the same way, it has been suggested that multipotent neural stem cells generate their progeny through a series of partially restricted intermediates, a process referred to as "neuropoiesis" (Fig. 1B; Anderson 1989).

While this overall concept is likely to be correct, the specifics of the logic and sequence of lineage restriction events remain unclear. For example, it has been widely assumed that neural stem cells would first generate restricted neuronal and glial progenitors, which would then in turn produce different types of neurons and different types of glia, respectively (Gage 1998: Fig. 1B). This scheme has a certain intuitive appeal. Furthermore, such neuron-restricted and glia-restricted progenitors cells have been defined in cultures derived from the embryonic spinal cord (Mayer-Proschel et al. 1997; Rao and Mayer-Proschel 1997). However, whether these intermediate progenitors reflect the normal pattern of cell fate-specification events as they occur in vivo has not yet been determined. In what follows, we will review evidence from studies of both the developing PNS and CNS that suggests a different sequence of fate-specification events, and we will discuss how this cellular logic can be rationalized in terms of the molecular mechanisms governing neural fate specification.

Molecular Mechanisms Governing Neural Cell Fate Specification

Before proceeding to discuss the cellular logic of neural lineage diversification, it is worth reviewing briefly what is known about the molecular mechanisms controlling neuronal and glial fate determination. These subjects have been reviewed in detail elsewhere (Jessell 2000; Bertrand et al. 2002).

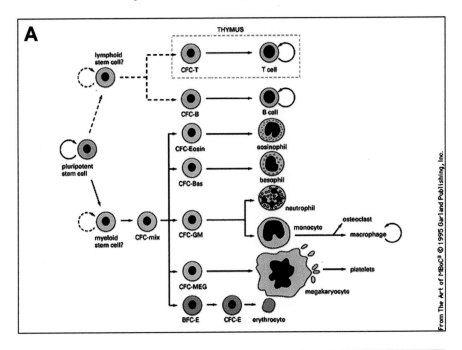

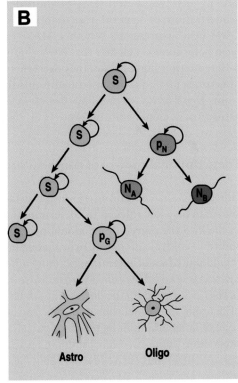

Fig. 1. Lineage diversification in the hematopoietic and nervous systems. **A)** Schematic diagram illustrating pattern of sequential restrictions during hematopoietic development. Reproduced with permission from Alberts et al. 1994. **B)** Conventional scheme for neural cell lineage diversification (Gage 1998). "p_N," committed neuronal progenitor; "p_G," committed glial progenitor. "N_A, N_B" indicate different subtypes of neurons.

Transcriptional control of the neuron vs. glial decision

Transcription factors in the basic helix-loop-helix (bHLH) family, called "proneural" genes, have emerged as central players in the decision between neuronal and glial fates. Genes of this type, including *achaete-scute* and *atonal*, were first identified in *Drosophila*, where they control the competence of the neuroectoderm to produce sensory organ precursors (SOPs) in the PNS and multipotent progenitors of neurons and glia ("neuroblasts") in the CNS (Jan and Jan 1994). In vertebrates, homologs of the genes have been identified, including *Mash1* (an *achaete-scute* homolog) and the *atonal*-related *Neurogenins* (reviewed in Bertrand et al. 2002). Loss- and gain-of-function studies have shown that these genes are both necessary and sufficient for neuronal differentiation, in both the PNS and CNS. As in *Drosophila*, different regions or lineages of the CNS and PNS employ different proneural genes. In the fly, this division of labor has been shown to reflect an additional role for these genes in the specification of neuronal subtype (Chien et al. 1996); whether this is also true in vertebrates has been controversial (see below).

In addition to their ability to promote neuronal differentiation, the neurogenins (NGNs) have also been shown to have an independent activity to actively inhibit glial (astrocyte) differentiation (Sun et al. 2001). This activity involves a non-DNA-binding–dependent mechanism that promotes sequestration of general transcriptional co-factors required for glial (i.e., *Gfap*) gene expression. Whether MASH1 also exhibits this activity has not been determined. The fact that NGNs both promote neuronal and inhibit glial differentiation suggests that they exert their function in multipotent progenitors of neurons and glia. Consistent with this finding, fetal progenitor cells isolated from *Ngn2-lacZ* –expressing knock-in mice generate both neurons and astrocytes in clonal culture (Nieto et al. 2001).

Transcriptional control of neuronal subtype specification

Homeodomain-containing transcription factors of various subclasses have been shown to play key roles in the determination of neuronal identity. For example, in the PNS the paired homeodomain factors PHOX2A and PHOX2B have been shown to control autonomic neuronal identity, in part via direct transcriptional regulation of catecholaminergic biosynthetic enzyme genes such as *tyrosine hydroxylase* and *dopamine β-hydroxylase* (reviewed in Goridis and Brunet 1999). In the CNS, studies in the developing spinal cord have shown that different types of neurons (motorneurons and interneurons) are specified by a combinatorial code of homeodomain factors that parcellate the ventricular zone into discrete progenitor domains, each producing a different neuronal subtype, along the dorso-ventral axis (Briscoe et al. 2000; Fig. 4, *left*). These homeodomain factors include genes such as *Nkx2.2, Irx3* and *Dbx1*. Other homeodomain factors control later steps in neuronal identity specification after progenitors have left the cell cycle. These include genes such as *Chx10*, which controls V2 interneuron identity, and *Mnr2/Hb9*, which controls motorneuron identity (Tanabe et al. 1998). Lim domain-containing homeodomain transcription factors also play an important role in specifying phenotypic features such as axon trajectory and neurotransmitter phenotype. These genes include *Lhx3, Lhx4, Isl1* and *Isl2* and appear to function in a combinatorial code as well (Pfaff and Kintner 1998).

An ongoing debate concerns the extent to which proneural genes control neuronal subtype identity as well as neuronal vs. glial differentiation (Hassan and Bellen 2000). As mentioned earlier, different neuronal subtypes depend on different proneural genes, and in *Drosophila* gain-of-function studies have shown that *achaete-scute* and *atonal* specify different subtypes of peripheral sensory organs (Jarman et al. 1993; Jarman and Ahmed 1998). Analogous studies in the vertebrate PNS have demonstrated that, similarly, MASH1 and NGN1/2 are sufficient to promote the differentiation of autonomic and sensory neurons, respectively, in cultured neural progenitor cells (Lo et al. 2002). However, in the case of the NGNs, this activity is far more sensitive to the influence of context than in the case of MASH1. Thus, for example, high concentrations of BMP2 can cause NGNs to promote an autonomic rather than sensory neuron identity, and substitution of *Ngn1* for *Mash1* coding sequences by homologous recombination rescues the differentiation of autonomic neurons caused by the deletion of *Mash1* (Parras et al. 2002). These data suggest that both MASH1 and the NGNs play some role in the specification of neuronal identity, but that in the case of the NGNs this role is more permissive than instructive. Nevertheless, the data suggest a functional interaction between proneural bHLH factors and neuronal identity-specification factors (Scardigli et al. 2001), and recent evidence suggests that this interaction may be a direct, physical one (S. Pfaff, personal communication).

Transcriptional control of glial subtype determination

Until recently, relatively little was known about the transcriptional control of glial fate determination in either the PNS or CNS. Sox10, an HMG-box-containing transcription factor, is required for the differentiation of all PNS glial cells (Britsch et al. 2001), but its role appears to be permissive rather than instructive (Paratore et al. 2001). Sox10 is also involved in oligodendrocyte differentiation in the CNS, but at a relatively late stage (Stolt et al. 2002). Transcriptional repressors in the bHLH family that are downstream effectors of Notch signaling, such as *Hes* and *Hey* genes, have been implicated in promoting astrocyte (Tanigaki et al. 2001) and Müller glial cell differentiation (Satow et al. 2001), respectively. However, whether these latter factors function primarily to promote glial differentiation, or rather to inhibit neuronal differentiation so that astrocyte differentiation may occur by default, is not yet certain. As discussed in more detail below, a new subfamily of bHLH factors, called *Olig* genes, has recently been shown to play a key role in the determination of the oligodendrocyte fate (reviewed in Rowitch et al. 2002). At least in part, this role involves the repression of astrocyte differentiation.

The Cellular Logic of Neural Cell Lineage Diversification

Sensory/autonomic neuronal fate specification precedes neuronal-glial fate specification in the PNS

The premigratory and early migrating neural crest contains multipotential cells that generate sensory and autonomic neurons, as well as glia (reviewed in Anderson 2000). Several lines of convergent evidence suggest that, in the neural crest, multipotent progenitors become fate-restricted with respect to the subtype(s) of neurons they generate, before they have become restricted to neuronal or glial fates. For example, multipotent, self-renewing rat neural crest stem cells (NCSCs) generate autonomic neurons of both the noradrenergic and cholinergic subtypes (Morrison et al. 2000), glial (Schwann) cells and smooth muscle/myofibroblast cells in vitro (Anderson 1997). When these cells are prospectively isolated from fetal peripheral nerve and transplanted into the neural crest migration pathway of chick embryos, they generate glia and neurons in various autonomic ganglia (Morrison et al. 1999); however they do not generate sensory neurons in dorsal root ganglia (DRG; White et al. 2001). In contrast, when a bulk population of rat neural crest cells (NCCs) is transplanted in the same assay, both autonomic neurons and sensory neurons are generated in the DRG (White and Anderson 1999). These data suggest that NCSCs may be restricted to generating autonomic neurons. Alternatively, the environment of the avian DRG may be permissive for rat sensory neuron differentiation from committed precursors present among NCCs, but not from multipotent NCSCs. Arguing in favor of a cell-instrinsic restriction in the sensory capacity of NCSCs is the fact that forced expression of NGN1 or NGN2 in these cells promotes autonomic but not sensory neuronal differentiation, although these same bHLH factors can promote sensory neurogenesis in other, more primitive, cell contexts (Lo et al. 2002).

These data suggest that the neural crest contains multipotent, self-renewing stem cells for neurons and glia (as well as smooth muscle) that are nevertheless restricted to generating autonomic neurons. Is the same true for progenitors of sensory neurons? To address this question, pre-migratory and early-migrating neural crest cells expressing *Ngn2* were fate-mapped using a hormone-inducible form of Cre recombinase to activate expression of the marker gene *lacZ* in these cells and their progeny (Zirlinger et al. 2002; Fig. 2A). These experiments revealed that *Ngn2*-derived cells were ~4 times more likely to contribute descendants to DRG than to autonomic (sympathetic ganglia), in comparison to bulk NCCs marked using the *Wnt1* promoter to drive hormone-inducible Cre (Danielian et al. 1998; Fig. 2B). Within DRG, however, *Ngn2*-derived cells, like *Wnt1*-derived cells, were equally likely to generate neurons and glia (Fig. 2C). Taken together with data from the CNS demonstrating that *Ngn*-expressing progenitors can clonogenically generate neurons and glia (Nieto et al. 2001), these data suggest that, in the neural crest, *Ngn2*-expressing cells become fate-restricted to generating sensory neurons before they are restricted to a neuronal fate. Whether these cells are irreversibly committed to a sensory fate, however, is not yet clear and will require prospective isolation and transplantation of such cells. Moreover, it is not known whether these cells are selfrenewing stem cells or non-self-renewing multipotent progenitors.

Taken together, these data suggest that, in the developing PNS, multipotent progenitors become fate-restricted to generating particular subtypes of neurons before they have become restricted to a neuronal fate (Fig. 3B). Such a pattern of lineage segregation appears inconsistent with the notion that multipotent neural stem cells first become restricted to neuronal or glial fates, and only later generate different subtypes of neurons and glia, respectively (Fig. 1B, 3A). However, in retrospect, this cellular logic makes sense in terms of the molecular mechanisms controlling neuron vs. glia and neuronal subtype determination. As mentioned earlier, vertebrate proneural genes control a neuronal vs. glial fate decision. Therefore, they must be expressed in progenitors before these cells have become committed to neuronal and glial fates, and the *Ngn2*- fate-mapping data are consistent with this theory (Nieto et al. 2001; Zirlinger et al. 2002). However, proneural genes also appear to collaborate with homeodomain neuronal identity factors to specify neuronal subtype (Lo et al. 1998, 2002; Scardigli et al. 2001;). This collaboration may involve both reciprocal functional and regulatory interactions (Scardigli et al. 2001). Such interactions imply that identity factors are co-expressed with proneural genes, which means they must also be expressed in progenitors that are not yet committed to neuronal or glial fates (Fig. 3C). Direct evidence for this expression has been obtained in the spinal cord (Novitch et al. 2001; Zhou et al. 2001). In this way, neural stem or progenitor cells would become restricted to particular neuronal subtype identities while they were still competent for neuronal and glial fates. Alternatively, such identity factors may not become expressed or brought into play until the proneural genes have committed multipotent cells to a neuronal fate. However, the transient nature of proneural gene expression argues that they exert their functions in a single progenitor cell compartment. In addition, data from the spinal cord provide further evidence for specification of neuronal subtype identity prior to the segregation of neuronal and glial lineages, as discussed below.

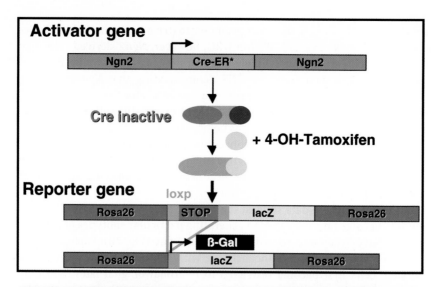

Activator gene

Ngn2 | Cre-ER* | Ngn2

Cre inactive

+ 4-OH-Tamoxifen

Reporter gene

loxp

Rosa26 | STOP | lacZ | Rosa26

ß-Gal

Rosa26 | lacZ | Rosa26

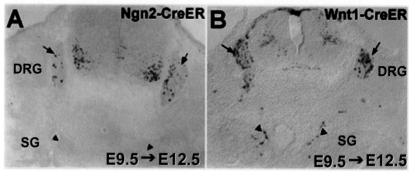

A Ngn2-CreER

DRG

SG

E9.5 → E12.5

B Wnt1-CreER

DRG

SG

E9.5 → E12.5

$DRG:SG_{Ngn2} \sim 20:1$ $DRG:SG_{Wnt1} \sim 5:1$

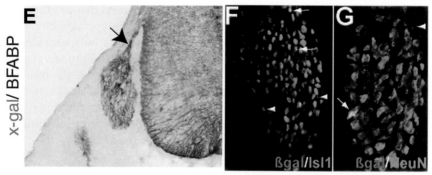

x-gal/ BFABP

E

F ßgal/Isl1

G ßgal/NeuN

$neuron:glia_{Ngn2} \sim 1:1$ $neuron:glia_{Wnt1} \sim 1:1$

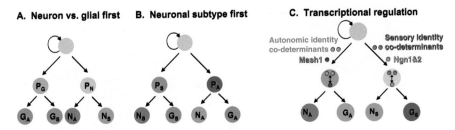

Fig. 3. Alternative models for neural lineage restriction in the PNS. **A)** A multipotent stem cell first generates committed glial (P_G) and neuronal (P_N) precursors, which generate autonomic and sensory glia (G_A and G_S) and autonomic and sensory neurons (NA and NS), respectively. **B)** The multipotent stem cell first generates sensory and autonomic multipotent precursors of neurons and glia (P_S and P_A, respectively). The data support Model B. **C)** Cross-regulation between proneural and neuronal identity genes may explain why identity specification precedes the neuron-glia decision.

Neuronal vs. glial subtype specification in the spinal cord

The ventricular zone of the spinal cord is parcellated into a series of discrete progenitor domains by the combinatorial expression of different homeodomain transcription factors (Briscoe et al. 2000; Fig. 4, *left*). Each domain generates a different subtype of neuron. In the ventral region, for example, the ventralmost p3 domain generates V3 interneurons, the overlying pMN domain generates motoneurons, and the p2 domain overlying pMN generates V2 interneurons (Fig. 4B). This spatial segregation of neuronal subtype specification occurs at stages of development long before any glia are generated from the spinal cord. Subsequently, oligodendrocytes and astrocytes are generated after the phase of neurogenesis is complete, at least in the ventral region (Fig. 4, *left*). Beginning on ~E12.5 in mouse, oligodendrocytes are generated from a restricted region of the ventricular zone, marked by expression of PDGFR-alpha (reviewed in Richardson et al. 1997; Richardson et al. 2000).

Expression of the bHLH transcription factor *Olig2* marks both the pMN domain, at the stage that motoneurons are being generated (Fig. 4A,B), and the oligodendrocyte progenitor domain at the stage that these myelinating glia are generated (Fig. 4D). Lossof-function studies in mice have shown that Olig2 is essential for both motoneuron and oligodendrocyte differentiation in vivo (Lu et al. 2002; Takebayashi et al. 2002; Zhou and Anderson 2002). Gain-of-

---

Fig. 2. Neurogenin2-expressing neural crest progenitors are biased to sensory but not neuronal fates. *Top*, schematic illustrating Cre-lox-dependent system for fate-mapping cells transiently expressing *Ngn2*. *Middle*, **A)** *Ngn2*-expressing cells labeled at E9.5 and assayed at E12.5 are predominantly located in DRG (arrows) and not sympathetic ganglia (SG, arrowheads) in the PNS. **B)** *Wnt1*-expressing cells, similarly labeled, populate both the DRG and SG. **E-G)** *Ngn2*-expressing cells generate neurons (F, G, arrows) and glia (E, arrow; F, G, arrowheads) in equal proportions. Modified from Zirlinger et al. 2002.

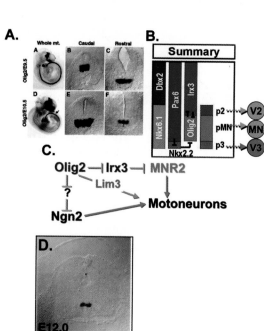

Fig. 4. Expression of *Olig2* defines the pMN and oligodendrocyte progenitor domains in the spinal cord. *Left*, schematic illustrating the parcellation of the spinal cord into progenitor domains that sequentially generate different types of neurons and glia. Modified with permission from Anderson 2001. *Right*, expression of *Olig2* defines the pMN domain at early stages (**A, B**) and promotes motoneuron differentiation in conjunction with *Ngn2* (**C**). **D)** Expression of *Olig2* in the oligodendrocyte progenitor domain at E12.0.

function studies in chick have shown that forced expression of *Olig2* alone, or in combination with *Ngn2*, generates ectopic motoneurons (Mizuguchi et al. 2001; Novitch et al., 2001; Fig. 4C), whereas forced expression of *Olig2* in combination with *Nkx2.2* generates premature and ectopic oligodendrocytes (Zhou et al. 2001). Taken together, these data demonstrate a key functional role for *Olig2* in the generation of both motoneurons and oligodendrocytes.

What is the fate of *Olig2*-expressing progenitor cells in *Olig2$^{-/-}$* mutants? Expression of a histone-EGFP fusion from the *Olig2* locus allowed short-term lineage tracing of *Olig2*- expressing cells in heterozygous and homozygous mice (Zhou and Anderson 2002). These studies revealed that, in the absence of *Olig2* function, presumptive motoneuron progenitors in the pMN domain adopted a V2 interneuron fate (Fig. 5a), while prospective oligodendrocyte progenitors adopted an astrocyte fate both in vivo and in vitro (Fig. 5b). These fate-transformations are consistent with the idea that *Olig2* couples a choice between alternative neuronal subtype identities (motoneuron vs. V2 interneuron) to a choice between alternative glial subtype identities (oligodendrocyte vs. astrocyte; (Fig. 5d; Zhou and Anderson, 2002). If, as suggested previously, the spinal cord ventricular zone contains a multipotent progenitor of motoneurons and oligodendrocytes (Richardson et al. 1997), these results would suggest that this progenitor is converted, in the absence of *Olig* gene function, to a common progenitor of V2 interneurons and astrocytes (Fig. 5c). In that case, in the CNS as in the PNS, multipotent progenitor cells would become fate-restricted to generating neurons and glia of a particular identity, before they became restricted to neuronal and glial fates (Fig. 6).

It has not yet been rigorously proven, however, that motoneurons and oligodendrocytes derive from a common clonogenic precursor. Therefore it remains formally possible that the pMN domain contains two separate precursor populations, both expressing *Olig2*: one population committed to producing only neurons (Mayer-Proschel et al. 1997), and the other committed to producing only glia (Rao and Mayer-Proschel 1997). In that case, these results would not necessarily imply that spinal cord ventricular zone progenitors become restricted to particular neuronal and glial subtype identities before they commit to neuronal and glial fates. However, lineage tracing studies in the spinal cord have demonstrated that motoneurons and glia share a common progenitor (although what kind of glia was not determined; Leber et al. 1990). Furthermore, at the stage that motoneurons are being generated (E9.5-E10.5), the majority of neuroepithelial precursor (NEP) cells in spinal cord cultures are multipotent (Kalyani et al. 1997). Therefore, the more parsimonious explanation of a common progenitor that depends sequentially on *Olig2* for motoneuron and then oligodendrocyte differentiation seems most likely (Kessaris et al. 2001).

If a common progenitor to motoneurons and oligodendrocytes exists, does it also produce astrocytes? In *Olig2^{hGFP/+}* heterozygous embryos, the EGFP marker is rarely if ever detected in astrocytes, implying that *Olig2*-expressing progenitors do not generate astrocytes (Zhou and Anderson 2002). While it is possible that the EGFP short-term lineage tracer is down-regulated below levels of detection before astrocytes are generated, similar experiments using a permanent lineage tracer expressed from the *Olig1* locus (which is co-expressed with *Olig2* in spinal cord) demonstrated labeling of motoneurons and oligodendrocytes, but not astrocytes (Lu et al. 2002).

Together, these data imply that the spinal cord contains two types of neuron-glia progenitor cells. One type expresses *Olig* genes and is fated to generate motoneurons and oligodendrocytes, but not astrocytes. A second type does not express *Olig* genes and is fated to generate interneurons and astrocytes, but not oligodendrocytes (Anderson et al. 2002). Are these multipotent progenitors stem cells? At this point, without isolating such cells and testing their self-renewal properties, this question cannot be answered. However, if a common clonogenic progenitor produces motoneurons and then oligodendrocytes in the pMN domain (or interneurons and then astrocytes in the p2 domain), then it would most likely have to divide asymmetrically to generate some daughter cells

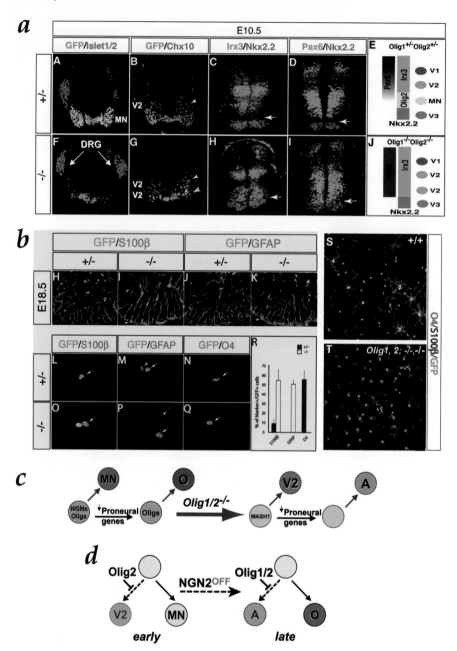

Fig. 5. *Olig2* couples neuronal and glial subtype identity. **a)** Marking of *Olig2*-expressing pro-genitors with GFP reveals that in the absence of *Olig2* (and *Olig1*) function these cells generate supernumerary V2 interneurons (G), instead of motoneurons (A). **b)** Similar embryos examined at later stages show that *Olig2*-expressing progenitor cells in heterozygotes generate oligodendrocytes in vivo (N) and in vitro (S), but not astrocytes (H, J, L, M, T). By contrast, in

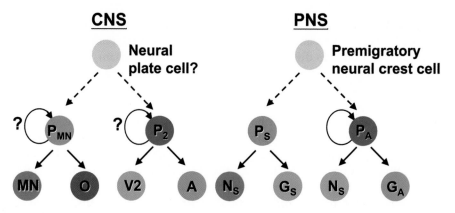

Fig. 6. Identity specification may precede neuronal and glial fate determination in both the CNS and PNS. Bipotent progenitors in the spinal cord (CNS) pMN domain generate motoneurons (MN) and oligodendrocytes (O), but not astrocytes, whereas those in p2 generate V2 interneurons and astrocytes (A), but not oligodendrocytes. Note that these neuronal and glial derivatives are generated sequentially, not simultaneously. It is not known whether these bipotent cells are self-renewing stem cells.

that remain undifferentiated during the neurogenic phase (Shen et al. 2002), in order to set aside sufficient cells to generate oligodendrocytes or astrocytes for the subsequent gliogenic phase. Such asymmetric divisions would be consistent with the contention, but not prove, that the cell is a self-renewing stem cell.

How does one square the idea that the spinal cord contains progenitor cells that generate neurons and astrocytes, or neurons and oligodendrocytes – but not all three cell types – with the definition of a CNS stem cell that has emerged from in vitro studies as a cell that generates neurons, astrocytes and oligodendrocytes (Temple 2001)? As mentioned earlier, one possibility is that the progenitors studied in vivo thus far are not stem cells but rather non-self-renewing progenitor cells (either multipotent or unipotent), and that the tripotent stem cells isolated in vitro correspond to an earlier stage of development. Another possibility is that the two different types of progenitor cells in vivo, although *fated* to generate either astrocytes or oligodendrocytes (but not both), have the *potential* to generate both types of glia, and this potential is revealed in vitro. In that case there would be no inherent contradiction between the in vivo and in vitro studies, although it would still be important to be aware of the difference between the cells' fate and their potential (Anderson 2001). However, it is possible that the tripotentiality revealed in vitro reflects an artfactual reprogramming of

---

homozygous mutants these cells generate astrocytes (I, K, O, P, T). **c)** Schematic illustrating the effect of the *Olig1/2* mutation on the sequential generation of motoneurons and astrocytes. **d)** *Olig2* is hypothesized to play independent roles in repressing V2 interneuron and astrocyte differentiation. Modified with permission from Zhou and Anderson 2002.

the cells, due to the tissue culture environment. To resolve this issue, it would be necessary to directly isolate the two types of progenitor cells (Olig2$^+$ and Olig2$^-$) and transplant them to different regions of the nervous system, to see whether they can each generate both classes of glia without being expanded ex vivo. It would also be necessary to test these cells for self-renewal and multipotency at the single-cell level.

In summary, evidence from both the PNS and CNS is consistent with the idea that the cellular logic of neural cell lineage diversification is one in which information specifying different neuronal or glial subtype identities is imposed on multipotent progenitor cells, before they have become restricted to neuronal or glial fates (Fig. 6). In at least one case, that of the NCSC, such neuronal subtype-restricted progenitors behave as self-renewing stem cells, at least in vitro (Stemple and Anderson 1992). The seemingly counter-intuitive notion that neuronal subtype identity is determined before progenitors have committed to a neuronal fate can be rationalized in terms of the regulatory interactions between proneural genes, which control the neuronal vs. glial fate choice, and homeodomain identity factors that specify different neuronal subtypes (Scardigli et al. 2001; Bertrand et al. 2002). It is also reflected in the spatial regionalization of the CNS ventricular zone, which ensures that different neuronal subtypes are generated in distinct regions of the developing CNS (Jessell 2000). These same spatial regionalization mechanisms also serve to segregate the generation of oligodendrocytes and astrocytes to different regions of the neuroepithelium. However, these patterns of fate restriction in vivo do not necessarily imply an equivalent restriction in developmental potential, which must be tested through transplantation and/or cell culture manipulations.

Acknowledgements

The authors thank Drs. Tom Jessell, Ben Novitch, Charles Stiles and David Rowitch for helpful advice, reagents and the free exchange of information throughout the course of these experiments. D.J.A. is an Investigator of the Howard Hughes Medical Institute. This work was supported in part by NIH grant NS23476.

References

Alberts B, Bray D, Lewis J, Raff M, Roberts K, Watson JD (1994) Molecular biology of the cell. New York & London, Garland Publishing, Inc.

Anderson DJ (1989) The neural crest cell lineage problem: Neuropoiesis? Neuron 3: 1–12

Anderson DJ (1997) Cellular and molecular biology of neural crest cell lineage determination. Trends Genet 13: 276–280

Anderson DJ (2000) Genes, lineages and the neural crest: a speculative review. Philos Trans R Soc Lond B Biol Sci 355: 953–964

Anderson DJ (2001) Stem cells and pattern formation in the nervous system: the possible versus the actual. Neuron 30: 19–35

Anderson DJ, Choi G, Zhou Q (2002) *Olig* genes and the genetic logic of CNS neural cell fate determination. Clin Neurosci Res 2: 17–28

Bertrand N, Castro DS, Guillemot F (2002) Proneural genes and the specification of neural cell types. Nature Rev Neurosci 3: 517–530

Briscoe J, Pierani A, Jessell TM, Ericson J (2000) A homeodomain protein code specifies progenitor cell identity and neuronal fate in the ventral neural tube. Cell 101: 435–445

Britsch S, Goerich DE, Riethmacher D, Peirano RI, Rossner M, Nave KA, Birchmeier C, Wegner M (2001) The transcription factor Sox10 is a key regulator of peripheral glial development. Genes Dev 15: 66–78

Chien C-T, Hsiao C-D, Jan LY, Jan YN (1996) Neuronal type information encoded in the basic-helix-loop-helix domain of proneural genes. Proc Natl Acad Sci USA 93: 13239–13244

Danielian PS, Muccino D, Rowitch DH, Michael SK, McMahon AP (1998) Modification of gene activity in mouse embryos in utero by a tamoxifen-inducible form of Cre recombinase. Curr Biol 8: 1323–1326

Gage F (2000) Mammalian neural stem cells. Science 287: 1433–1438

Gage FH (1998) Stem cells of the central nervous system. Curr Opin Neurol 8: 671–676

Goridis C, Brunet JF (1999) Transcriptional control of neurotransmitter phenotype. Curr Opin Neurobiol 9: 47–53

Hassan BA, Bellen HJ (2000) Doing the MATH: is the mouse a good model for fly development? Genes Dev 14: 1852–1865

Hitoshi S, Tropepe V, Ekker M, van der Kooy D (2002) Neural stem cell lineages are regionally specified, but not committed, within distinct compartments of the developing brain. Development 129: 233–244

Jan YN, Jan LY (1994) Genetic control of cell fate specification in the Drosophila peripheral nervous system. Ann Rev Genet 28: 373–393

Jarman AP, Ahmed I (1998) The specificity of proneural genes in determining Drosophila sense organ identity. Mech Dev 76: 117–125

Jarman AP, Grau Y, Jan LY, Jan Y-N (1993) *atonal* is a proneural gene that directs chordotonal organ formation in the Drosophila peripheral nervous system. Cell 73: 1307–1321

Jessell TM (2000) Neuronal specification in the spinal cord: inductive signals and transcriptional codes. Nature Rev Genet 1: 20–29

Kalyani A, Hobson K, Rao MS (1997) Neuroepithelial stem cells from the embryonic spinal cord: isolation, characterization, and clonal analysis. Dev Biol 186:202–223

Kessaris N, Pringle N, Richardson WD (2001) Ventral neurogenesis and the neuron-glial switch. Neuron 31: 677–680

Leber SM, Breedlove SM, Sanes JR (1990) Lineage, arrangement, and death of clonally related motorneurons in chick spinal cord. J Neurosci 10: 2451–2462

Lo L, Tiveron M-C, Anderson DJ (1998) MASH1 activates expression of the paired homeodomain transcription factor Phox2a, and couples pan-neuronal and subtype-specific components of autonomic neuronal identity. Development 125: 609–620

Lo L, Dormand E, Greenwood A, Anderson DJ (2002) Comparison of the generic neuronal differentiation and neuron subtype specification functions of mammalian achaete-scute and atonal homologs in cultured neural progenitor cells. Development 129: 1553–1567

Lu QR, Sun T, Zhu Z, Ma N, Garcia M, Stiles CD, Rowitch DH (2002) Common developmental requirement for Olig function indicates a motor neuron/oligodendrocyte connection. Cell 109: 75–86

Mayer-Proschel M, Kalyani AJ, Mujtaba T, Rao MS (1997) Isolation of lineage-restricted neuronal precursors from multipotent neuroepithelial stem cells. Neuron 19: 773–785

Mizuguchi R, Sugimori M, Takebayashi H, Kosako H, Nagao M, Yoshida S, Nabeshima Y, Shimamura K, Nakafuku M (2001) Combinatorial roles of olig2 and neurogenin2 in the coordinated induction of pan-neuronal and subtype-specific properties of motoneurons. Neuron 31: 757–771

Morrison SJ, Csete M, Groves AK, Melega W, Wold B, Anderson DJ (2000) Culture in reduced levels of oxygen promotes clonogenic sympathoadrenal differentiation by isolated neural crest stem cells. J Neurosci 20: 7370–7376

Morrison SJ, White PM, Zock C, Anderson DJ (1999) Prospective identification, isolation by flow cytometry, and in vivo self-renewal of multipotent mammalian neural crest stem cells. Cell 96: 737–749

Nieto M, Schuurmans C, Britz O, Guillemot F (2001) Neural bHLH genes control the neuronal versus glial fate decision in cortical progenitors. Neuron 29: 401–413

Novitch BG, Chen AI, Jessell TM (2001) Coordinate regulation of motor neuron subtype identity and pan-neuronal properties by the bHLH repressor Olig2. Neuron 31: 773–789

Paratore C, Goerich DE, Suter U, Wegner M, Sommer L (2001) Survival and glial fate acquisition of neural crest cells are regulated by an interplay between the transcription factor Sox10 and extrinsic combintorial signaling. Development 128: 3949–3961

Parras CM, Schuurmans C, Scardigli R, Kim J, Anderson DJ, Guillemot F (2002) Divergent functions of the proneural genes Mash1 and Ngn2 in the specification of neuronal subtype identity. Genes Dev 16: 324–338

Pfaff S, Kintner C (1998). Neuronal diversification: development of motor neuron subtypes. Curr Opin Neurobiol 8: 27–36

Rao MS, Mayer-Proschel M (1997) Glial-restricted precursors are derived from multipotent neuroepithelial stem cells. Dev Biol 188: 48–63

Richardson WD, Pringle NP, Yu W-P, Hall AC (1997) Origins of spinal cord oligodendrocytes: possible developmental and evolutionary relationships with motor neurons. Dev Neurosci 19: 58–68

Richardson WD, Smith HK, Sun T, Pringle NP, Hall A, Woodruff R (2000) Oligodendrocyte lineage and the motor neuron connection. Glia 29: 136–142

Rowitch DH, Lu QR, Kessaris N, Richardson WD (2002) An 'oligarchy' rules neural development. Trends Neurosci 25: 417–422

Satow T, Bae S, Inoue T, Inoue C, Miyoshi G, Tomita K, Bessho Y, Hashimoto N, Kageyama R (2001) The basic Helix-Loop-Helix gene Hesr2 promotes gliogenesis in mouse retina. J Neurosci21: 1265–1273

Scardigli R, Schuurmans C, Gradwohl G, Guillemot F (2001) Crossregulation between *Neurogenin2* and pathways specifying neuronal identity in the spinal cord. Neuron 31: 203–217

Scherer DC, Miyamoto T, King, AG, Akashi K, Sugamura K, Weissman IL (2000) A clonogenic common myeloid progenitor that gives rise to all myeloid lineages. Nature 407: 383–386

Shen Q, Zhong W, Jan YN, Temple S (2002) Asymmetric Numb distribution is critical for asymmetric cell division of mouse cerebral cortical stem cells and neuroblasts. Development 129: 4843–4853

Stemple DL, Anderson DJ (1992) Isolation of a stem cell for neurons and glia from the mammalian neural crest. Cell 71: 973–985

Stolt CC, Rehberg S, Ader M, Lommes P, Riethmacher D, Schachner M, Bartsch U, Wegner M (2002) Terminal differentiation of myelin-forming oligodendrocytes depends on the transcription factor Sox10. Genes Dev16: 165–170

Sun Y, Nadal-Vincens M, Misono S, Lin MZ, Zubiaga A, Hua X, Fan G, Greenberg ME (2001) Neurogenin promotes neurogenesis and inhibits glial differentiation by independent mechanisms. Cell 104: 365–376

Takebayashi H, Nabeshima Y, Yoshida S, Chisaka O, Ikenaka K (2002) The basic helix-loop-helix factor olig2 is essential for the development of motoneuron and oligodendrocyte lineages. Curr Biol 12: 1157–1163

Tanabe Y, William C, Jessell TM (1998) Specification of motor neuron identity by the MNR2 homeodomain protein. Cell 95: 67–80

Tanigaki K, Nogaki F, Takahashi J, Tashiro K, Kurooka H, Honjo T (2001) Notch1 and Notch3 instructively restrict bFGF-responsive multipotent neural progenitor cells to an astroglial fate. Neuron 29: 45–55

Temple S (2001) The development of neural stem cells. Nature 414: 112–117

White PM, Anderson DJ (1999) In vivo transplantation of mammalian neural crest cells into chick hosts reveals a new autonomic sublineage restriction. Development 126: 4351–4363

White PM, Morrison SJ, Orimoto K, Kubu CJ, Verdi JM, Anderson DJ (2001) Neural crest stem cells undergo cell-intrinsic developmental changes in sensitivity to instructive differentiation signals. Neuron 29: 57–71

Zhou Q, Anderson DJ (2002) The bHLH transcription factors OLIG2 and OLIG1 couple neuronal and glial subtype specification. Cell 109: 61–73

Zhou Q, Choi G, Anderson DJ (2001) The bHLH transcription factor Olig2 promotes oligodendrocyte differentiation in collaboration with Nkx2.2. Neuron 31:791–807

Zirlinger M, Lo L, McMahon J, McMahon AP, Anderson DJ (2002) Transient expression of the bHLH factor neurogenin-2 marks a subpopulation of neural crest cells biased for a sensory but not a neuronal fate. Proc Natl Acad Sci USA 99: 8084–8089

Astrocytic nature of adult neural stem cells in vivo

A. Alvarez-Buylla[1], F. Doetsch[1], B. Seri[1] and *J.M. Garcia-Verdugo[1]*

Summary

The identification of adult brain regions harboring neural stem cells and of their continual generation of new neurons throughout life challenges traditional views of the adult brain's germinal potential and suggests new alternatives for brain repair. Surprisingly, cells with characteristics of mature astrocytes have been shown to function as the primary precursors of new neurons in vivo. This observation goes against the generalized assumption that the rare cells with undifferentiated appearance are neural stem cells. It suggests that neural stem cells in vivo have important structural functions in addition to their role as primary progenitors. Surprisingly, neurospheres – a commonly used model of neural stem cells in culture – are not necessarily derived from the in vivo primary precursors but from transit amplifying cells. We suggest that neural stem cells, when activated to become transit amplifying cells, upregulate the expression of growth factor receptors on their surface, allowing their isolation as neurospheres.

Introduction

Neurogenesis and the recruitment of new neurons into functional circuits have been shown to occur in the telencephalon of adult birds (Goldman and Nottebohm 1983) and in restricted regions of the adult mammalian forebrain (Altman 1970). These findings suggest that germinal centers persist in the adult CNS. How these germinal centers are organized and which cells within them function as primary precursors or stem cells has been the focus of our research. Here we discuss data from studies in adult birds and mammals that indicate that primary precursors have characteristics of cells that were considered for many years exclusively as part of the astroglial lineage.

[1] Department of Neurological Surgery, Brain Tumor Research Center, Box 0520, 533 Parnassus Ave., San Francisco, CA 94143-0520, Tel: (415) 514-2348, Fax: (415) 514-2346, e-mail: abuylla@itsa.ucsf.edu

Gage et al.
Stem Cells in the Nervous System:
Functional and Clinical Implications
© Springer-Verlag Berlin Heidelberg 2004

In adult song birds, new neurons are incorporated into multiple telencephalic circuits, including those controlling song (Goldman and Nottebohm 1983; Paton and Nottebohm 1984; Nottebohm 1985; Alvarez-Buylla and Nottebohm 1988). Neurons are born in the walls of the lateral ventricles of the adult avian brain (Goldman and Nottebohm 1983; Barami et al. 1995) and migrate long distances to reach several areas of the telencephalon (Alvarez-Buylla and Nottebohm 1988). New neurons continuously replace old ones (Kirn and Nottebohm 1993; Nottebohm and Alvarez-Buylla 1993) in a process thought to be related to plasticity and learning (Nottebohm 1985; Alvarez-Buylla et al. 1990a).

Neuronal birth in adult mammals has been demonstrated in two brain regions:

1. the subventricular zone (SVZ; Lois and Alvarez-Buylla 1993, 1994), and
2. the subgranular layer (SGL) of the hippocampal dentate gyrus (Kaplan and Bell 1984; Cameron et al.1993; Gage et al. 1998).

In the SVZ, cells are born over a large area adjacent to the lateral ventricle. From their site of birth, chains of young neurons migrate along a complex network of tangential pathways and join the rostral migratory stream (RMS), which leads to the olfactory bulb, where new cells differentiate into two types of local interneurons, granule and periglomerular neurons (Lois and Alvarez-Buylla 1994; Doetsch and Alvarez-Buylla 1996). Cells born in the SGL migrate a short distance and differentiate into new granule neurons within the dentate gyrus (Kaplan and Hinds 1977; Bayer et al. 1982; Cameron et al. 1993; Kuhn et al. 1996)

Identification of primary precursors of new neurons in adult birds and mammals

Proliferative activity in the adult brain has been classically associated with the renewal of glial cells. It was believed that the embryonic ventricular zone (VZ), as a structure, and the neuron-producing stem cells therein disappeared soon after birth. We now know that this is not the case. Germinal layers persist in the adult brain, suggesting that their stem cells could be used for brain repair. Thus the identification of the primary precursors in the adult brain and the organization of the germinal layers that harbor them are of considerable interest.

A VZ in the adult avian brain. The proliferative layer in the walls of the lateral ventricles in adult birds has similar properties to the developing VZ (Alvarez-Buylla et al. 1998). It is in this VZ that the neural stem cells that support adult neurogenesis in birds reside. Proliferation occurs at higher rates in VZ "hot spots," where a large number of new neurons are born (Alvarez-Buylla et al. 1990b). These hot spots contain a large number of radial glia (Alvarez-Buylla et al. 1987). Radial glia have been studied extensively in the embryonic brain of mammals, as they serve as scaffolding for building the brain (Rakic 1988) by guiding the migration of young neurons (Rakic 1972). In mammals, radial glia were considered committed progenitors of parenchymal astrocytes (Ramon y Cajal 1911; Schmechel and Rakic 1979; Voigt 1989).

As in mammals, the long shaft of radial glia in adult birds guides the migration of young neurons (Alvarez-Buylla and Nottebohm 1988). Very few glial cells, if any, are formed in the VZ of adult birds (Alvarez-Buylla and Nottebohm 1988). It was therefore surprising to find that, in the adult avian brain, radial glia divide. Interestingly, their division correlates spatially and temporally with the appearance of new neurons, leading to the conclusion that radial cells in adult songbirds divide to generate new neurons (Alvarez-Buylla et al. 1990b). Retroviral lineage studies in the adult avian telencephalon are also consistent with the interpretation that radial glia function as primary neuronal precursors (Goldman 1995). The above data suggest that radial glia possess properties of neuroepithelial progenitors, and their differentiation potential is not limited to the generation of glial cells. Interestingly, radial glia in adult birds have an elongated morphology and contact both the pial and ventricular surface of the brain, similar to neuroepithelial cells. Both radial glia and neuroepithelial cells also possess a single short cilium that extends into the cerebrospinal fluid (Cohen and Meininger 1987; Alvarez-Buylla et al. 1998). The function of this cilium is not known. The VZ of adult birds also contains multiciliated ependymal cells, but these cells do not divide.

The proposition that radial glia could function as primary neuronal precursors was initially received with skepticism, as previous work in mammals pointed to radial glia as precursors of astrocytes only (Levitt et al. 1981; Voigt 1989). However, more recent evidence has demonstrated that radial glia do function as neuronal progenitors in the developing mammalian brain (Malatesta et al. 2000; Miyata et al. 2001; Noctor et al. 2001). As in birds, radial glia have a complex structure far from what could be catalogued as an undifferentiated cell. This complex structure is likely related to the many functions that these cells play in addition to being progenitors. Their supporting role as scaffolding during development is well established (Rakic 1988).

Astrocytes as progenitors in the adult SVZ. The presence of large numbers of dividing cells in the SVZ has been known for almost a century (Allen 1912) and is known to occur in many adult vertebrates (Lewis,1968; Blakemore and Jolly 1972; McDermott and Lantos 1990; Huang et al. 1998; Gould et al. 1999; Kornack and Rakic 2001), including humans (Eriksson et al. 1998). However, the nature and fate of newly generated SVZ cells have been controversial. Most studies suggested that dividing SVZ cells gave rise to new glial cells, (e.g. Smart and Leblond 1961; Privat and Leblond 1972; Privat 1977). Recent studies are consistent with the proposition that glial cells are born in the adult SVZ (Levison and Goldman 1993; Goldman 1995; Nait-Oumesmar et al. 1999). Other investigators suggested that dividing cells in the SVZ die soon after mitosis

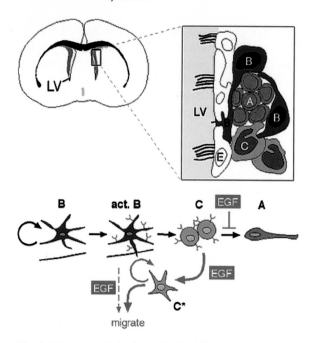

Fig. 1. Neurogenesis in the SVZ of adult mice. Schematic of a frontal section (upper left) through the anterior adult mouse forebrain, indicating the location of the SVZ (dark gray) next to the lateral ventricles. The SVZ organization and composition are shown in the box to the right. The SVZ is separated from the lateral ventricle by a layer of multiciliated ependymal cells (E, white). Chains of neuroblasts (A, red) travel through glial tunnels formed by the processes of SVZ astrocytes (B, blue). Clusters of rapidly dividing transit-amplifying C cells (green) are scattered along the network of chains. Occasionally, the process of an SVZ astrocyte comes into contact with the lateral ventricle. These SVZ astrocytes have a short single cilium. SVZ astrocytes (B, blue) act as stem cells in this region and divide to generate transit-amplifying C cells (green), which in turn divide to generate neuroblasts (A, red) that migrate to the olfactory bulb. The normal lineage of SVZ stem cells (black arrow) and the effects of EGF (in red) are depicted at the bottom of the figure. EGF infusion blocks the generation of A cells and induces C cells, and probably activated astrocytes (act. B), to divide and become highly invasive.

(Smart 1961; Morshead and Van der Kooy 1992). However, most of the dividing cells correspond to neuronal progenitors (Lois and Alvarez-Buylla 1993). As indicated above, these neurons migrate to the olfactory bulb, where they constantly replace interneurons (Luskin 1993; Lois and Alvarez-Buylla 1994), a process that also continues in adult primates (Kornack and Rakic 2001). It has been suggested that SVZ cells may give rise to neurons that migrate to other brain regions in primates (Gouldet al. 1999; Magavi et al. 2000), but this process remains to be clearly demonstrated.

The cellular composition and organization of the SVZ have been described using immunocytochemistry and electron microscopy (Doetsch et al. 1997;

Fig. 1). SVZ neuroblasts (type A cells) are organized as homotypic chains (Lois et al. 1996; Wichterle et al. 1997). These chains are pathways of migration distributed throughout the wall of the lateral ventricle, as described above (Doetsch and Alvarez-Buylla 1996), and are ensheathed by slowly proliferating type B cells (Lois et al. 1996). B cells have properties of astrocytes, including a light cytoplasm, thick bundles of glial fibrillary acidic protein (GFAP)-positive intermediate filaments, gap junctions and glycogen granules (Doetsch et al. 1997). Scattered along the chains of type A cells are clusters of rapidly dividing type C cells (Doetsch et al. 1997). The SVZ is largely separated from the ventricle by a layer of multiciliated ependymal cells, but the chains of migrating cells do not contact the ependymal cells directly. Instead B cells extend thin protrusions that separate chains of A cells from the ependymal layer. Many of the chains of type A cells coalesce in the anterior and dorsal SVZ, forming the RMS (Lois and Alvarez-Buylla 1994).

SVZ astrocytes function as the primary neuronal precursors in vivo. Following ablation of actively dividing cells by infusion of the antimitotic drug cytosine-β-D-arabinofuranoside (Ara-C) into the adult mouse brain for six days, neuroblasts and rapidly dividing C cells are eliminated. This treatment spares some SVZ astrocytes and ependymal cells (Doetsch et al. 1999a). Twelve hours after Ara-C removal, SVZ astrocytes begin dividing. Dividing SVZ astrocytes give rise to C cells, which in turn generate neuroblasts (Doetsch et al. 1999b). Within 10 days, the entire SVZ regenerates. Ependymal cells do not incorporate mitotic markers at any of the survival times studied, indicating that these cells are not the primary precursors of the new neurons.

Consistent with the above results, GFAP-expressing astrocytes under normal conditions are primary precursors for new neurons. Specific infection of astrocytes in transgenic mice carrying the receptor for an avian retrovirus under the GFAP promoter generate neurons that migrate to the olfactory bulb, indicating that under normal conditions SVZ astrocytes are the primary precursors of new neurons (Holland and Varmus 1998; Doetsch et al. 1999a). Interestingly, astrocytes from brain regions outside the SVZ can also generate multipotent neurospheres, but only if isolated from mice younger than 10 days (Laywell et al. 2000). After this time, only astrocytes from the SVZ have been shown to have this capacity (Laywell et al. 2000). The above results suggest that neural stem cells correspond to cells that are classically considered differentiated macroglia (Privat and Leblond 1972).

Multiciliated ependymal cells have also been suggested to function as SVZ neural stem cells (Johansson et al. 1999), but this suggestion is not supported by other studies (Chiasson et al. 1999; Doetsch et al. 1999a; Laywell et al. 2000; Irvin et al. 2001). There is no evidence that multiciliated ependymal cells divide in vivo (Alvarez-Buylla et al. 1998; Doetsch et al. 1999a). Ependymal cells do, however, play an important role in adult neurogenesis (Lim et al. 2000). The neurogenic potential of many SVZ astrocytes is under constant inhibition by bone morphogenetic proteins (BMPs). Local noggin production, which neutralizes the inhibitory effects of BMPs in the SVZ, has been suggested to release this inhibition and allow astrocytic precursors to become neurogenic. Ependymal cells (Lim et al. 2000) express noggin, which binds BMPs and neutralizes its inhibitory actions. Alternatively (or in addition), astrocytes themselves appear to express noggin (unpublished observation), although at lower concentrations than ependymal cells, and these cells may also contribute to the neutralization of the BMPs inhibition. However, it is not known how noggin expression is locally controlled and what determines the activation of some astrocytes to produce neurons. The process has to be tightly regulated to prevent neuroblast overproduction. SVZ astrocytes make extensive contacts with ependymal cells and occasionally extend a process that reaches the ventricle. Their close contact with ependymal cells and contact with the ventricle may be important for the local activation of a subpopulation of the astrocytes in the SVZ.

How the proliferation of SVZ astrocytes is regulated, and what factors activate them to function as progenitors of neurons, is not known. The number of dividing cells in the SVZ is affected by lesions (Szele and Chesselet 1996; Nait-Oumesmar et al. 1999), activation of Eph receptors (Conover et al. 2000), inhibition of nitric oxide (Moreno-Lopez et al. 2000), estrogen (Smith et al. 2001) and prolactin during pregnancy (Shingo et al. 2003). However, it is not known which of these treatments specifically affects the primary precursors. Most of the dividing cells in the SVZ correspond to C cells, and it is likely that the above treatments affect this population. Quantification of proliferation is further complicated since tangentially migrating young neurons also divide (Lois and Alvarez-Buylla. 1994).

In addition, we do not know which astrocytes function as stem cells, since astrocytes in the brain are very heterogeneous in structure and function. More specific markers are required to identify the different classes of astrocytes in the adult brain and in particular those that can function as stem cells. As mentioned above, neurogenesis also continues in the SGL layer of the hippocampus. In the next section we will review recent work that suggests that astrocytes in the SGL also function as the primary precursors for new neurons.

Astrocytes as progenitors in the adult SGL. As indicated above, the SGL is the site of birth of new neurons in the dentate gyrus of the hippocampus. Unlike the SVZ, the SGL is not located close to the walls of the brain ventricles, but deep within the hippocampus at the interface of the hilus and the granule cell layer. In vitro studies have suggested that the adult hippocampus contains neural stem cells (Palmer et al. 1997; Gage 2000), but it is not clear that these cells are related to the progenitors of new neurons in vivo. Earlier in vivo studies identified small dark cells in the SGL as neuronal progenitors (Altman and Das 1965; Kaplan and Bell 1984). Other studies have shown that astrocytes continue to divide in the SGL of the adult hippocampus (Cameron et al. 1993; Palmer et al. 2000), but this process was attributed to local gliogenesis necessary for the maintenance and support of neuronal function. Given the evidence showing that SVZ-astrocytes give rise to new neurons, we investigated the possibility that SGL astrocytes could function as primary neuronal precursors (Seri et al. 2001).

A subpopulation of SGL astrocytes stain prominently with antibodies to GFAP: their cell bodies are in the SGL, and they extend a radial process into the granule cell layer and short tangential processes along the blades of the dentate gyrus (Kosaka and Hama 1986). These cells have ultrastructural features of astrocytes (Seri et al. 2001). Interestingly, soon after a BrdU (bromodeoxyuridine) or a [^{3}H]-Thymidine injection, 60–80% of labeled cells in the SGL correspond to these astrocytes, but their number decreases rapidly, coinciding with an increase in the number of labeled GFAP-negative cells with similar characteristics to the small dark cells (D cells) described by others (Altman and Das 1965; Kaplan and Bell 1984). D cells are rarely, if ever, found alone. Instead, they are always tightly associated with SGL astrocytes and sometimes ensheathed by them.

Three lines of evidence indicate that dividing SGL astrocytes serve as primary precursors for new neurons (Seri et al. 2001): 1) following elimination of dividing cells (including D cells) in the SGL, dividing astrocytes regenerate granule neurons; 2) selective infection of SGL astrocytes using the RCAS-AP avian retrovirus in transgenic mice (Holland and Varmus 1998), results in AP-labeled granule neurons one month later; 3) proliferating label-retaining cells in the SGL, one month after [^{3}H]-thymidine administration, are not D cells but correspond to astrocytes. Together, these experiments indicate that SGL astrocytes give rise to new granule neurons in the adult dentate gyrus through the intermediate D cells (Fig. 2).

The radial organization of SGL astrocytes that function as neuronal progenitors suggests that these cells could also be involved in the regulation of neurogenesis. Neurogenesis and/or neuronal recruitment in the adult dentate gyrus can be altered by adrenal steroids (Gould et al. 1992), the blockage of NMDA receptors (Gould et al. 1994), seizures (Parent et al. 1997), housing animals in enriched conditions (Kempermann et al. 1997) or exercise (Van Praag et al. 1999). The way in which these different treatments affect proliferation of SGL progenitors is not known. Somehow progenitor proliferation/survival at the level of the SGL has to be altered, depending on demands at the level of the granule cell layer. Radial SGL astrocytes are well suited for such function; they act as primary progenitors, and at the same time have a long radial process that traverses the granule cell layer. Interestingly, neuronal birth in the dentate gyrus increases following adrenalectomy (Gould et al. 1992), and this same treatment increases the proliferation of astrocytes. Astrocytes may become activated to accommodate the demand for more neurons following adrenalectomy. The radial process of SGL astrocytes may serve as a sensor of signals, within the granule cell layer and perhaps the molecular layer, that affect neurogenesis in the dentate gyrus (Seri and Alvarez-Buylla 2002).

In addition to their role as progenitors and as possible regulatory signals transducers, astrocytes may also provide important support for the

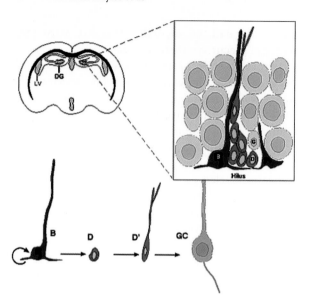

Fig. 2. Neurogenesis in the hippocampus SGL of adult mice. Schematic of a frontal section through the posterior adult mouse forebrain, indicating the location of the dentate gyrus (DG) in the hippocampus (upper left). In the box to the right, the composition and architecture of the SGL and granule cell layer are indicated. SGL radial astrocytes (B cells) have long processes that penetrate the granule cell layer and short tangential ones that run parallel to the blades of the dentate gyrus. B cells are tightly associated with small dark cells (D cells) forming clusters in the SGL. The processes of B cells ensheath the clusters of D cells, insulating them from the hilus. Dividing B cells apparently generate D cells, which in turn differentiate into new granule neurons (G). D cells are initially small and round but then develop a process (D'), that matures into the apical dendrite of the new neuron. GC, granule cell.

differentiation and maturation of new neurons. As first shown for the SVZ, efficient neuronal formation in vitro requires direct contact with astrocytes (Lim and Alvarez-Buylla 1999). A similar phenomenon has been observed with expanded neural stem cells and hippocampal astrocytes (Song et al. 2002). Thus, at least a population of SGL astrocytes may be multifunctional cells mothering and tending their own progeny.

Transit-amplifying cells give rise to neurospheres. It has been argued that stem cells, defined as self-renewing multipotent cells, reside in germinal regions of the adult. The extent to which primary precursors self-renew in vivo is not known, but in vitro studies suggest that cells exposed to high concentrations of EGF or bFGF can self-renew. These expanded cells retain the ability to differentiate into astrocytes, oligodendrocytes and neurons. Cells that possess the properties of stem cells in culture can be isolated from the SVZ using EGF, bFGF or both (Morshead et al. 1994; Gritti et al. 1999; Rietze et al. 2001). These expanded cells may be useful for cell replacement therapies (Weiss et al. 1996; McKay 1997; Gage

2000). Similar neural stem cells can also be obtained from other brain regions (Weiss et al. 1996; Palmer et al. 1999), but it is only from the SVZ that these in vitro stem cells can be isolated using EGF alone (Weiss et al. 1996). Given these properties, it has been suggested that cells that in vitro behave as stem cells are derived from rare and slowly dividing neural stem cells in vivo (Morshead et al. 1994).

In contrast, recent work from our laboratory has shown that the majority of the EGF-responsive cells in the SVZ are not derived from the primary precursor astrocytes, but from the actively dividing transit-amplifying progenitors, Type C cells (Doetsch et al. 2002a). Killing Type C cells in transgenic mice expressing HSV-TK under the promoter of DLX2, a homeobox gene expressed by C cells, results in approximately 70% reduction in the number of EGF-responsive neurospheres formed. In addition, purified Type C cells can generate neurospheres. Type C cells express EGF receptors in vivo and respond to EGF infusion by increasing their proliferation, halting their differentiation into neuroblasts and becoming highly invasive. Importantly, the neurospheres derived from C cells in vitro are multipotent and self-renewing, indicating that these transit-amplifying cells are not irreversibly committed to a differentiation pathway. Supra-physiological concentrations of growth factors and the extended proliferation of C cells may block them from their normal differentiation path.

Artificial expansion of transit-amplifying cells by growth factors, and in particular EGF, may have a biological basis. Signaling through the EGF receptor pathway may be part of the normal, regulated growth of C cells. After activation, SVZ astrocytes must engage in proliferation and transformation into C cells. This process is likely highly regulated in vivo by the balanced action of growth factor receptors and transcription factors on activated astrocytes and the transit-amplifying cells. In addition, growth-arresting molecules like p53 (Van Lookeren Campagne and Gill 1998), p27 (Doetsch et al. 2002b) and probably other molecules may serve as natural brakes in the amplification process, stopping the proliferation of C cells and allowing C cells to differentiate. Interestingly, mutations in the p27 gene result in a significant increase in the number of C cells in the SVZ. The interplay of stimulation, proliferation and accumulation of transcription factors may be essential for the balance between amplification of cell number and differentiation. This process is likely tightly regulated in vivo to prevent overgrowth. The source of growth factor/s that activate the EGF receptor in vivo in SVZ progenitors and its mode of regulation are not know. Interestingly, the absence of TGF-alpha, which also activates the EGF receptor, significantly reduces the number of proliferating cell in the SVZ (Tropepe et al. 1997). The supra-normal activation of EGF receptors by the addition of EGF in vitro or in vivo may drive transit-amplifying cells into continual expansion, preventing their normal conversion into neuroblasts. Infusion of EGF into the SVZ results in the expansion of proliferating cells in the SVZ and the almost complete arrest of neuroblast production (Craig et al. 1996; Kuhn et al. 1997; Doetsch et al. 2002a). C cells may be the culprit in the initiation of some types of brain tumors. Their invasive behavior after EGF infusion in vivo is reminiscent of that of invasive glioblastomas.

Conclusions

Cells previously considered to be part of the astroglial lineage have been shown to possess characteristics of primary progenitors in the embryonic and adult brain. Evidence from several laboratories has now shown that radial glial cells serve as precursors of neurons (Alvarez-Buylla et al. 1990; Malatesta et al. 2000; Noctor et al. 2001, 2002). Many studies have also shown that radial glial cells transform into astrocytes at the end of fetal development in mammals. As reviewed above, cells with the characteristics of astrocytes also function as neuronal precursors (Doetsch et al. 1999a; Laywell et al. 2000; Skogh et al. 2001; Seri et al. 2001). Ramon y Cajal suspected that astrocytes were nothing more than displaced and transformed neuroepithelial cells (1911). The progenitor potential of astrocytes and radial glia is likely derived from their ontogenetic and phylogenetic origin in neuroepithelial cells. Based on these observations, we suggest that early neuroepithelial cells, radial glia and astrocytes are the stem from which most brain cell lineages derive. Thus basic support cells act as precursors for terminally differentiated brain cells (Alvarez-Buylla et al. 2001). The abilities to generate multiple progenies and support their migration and maturation are two essential functions that radial and astroglial neural stem cells may have inherited from primitive neuroepithelial cells.

Acknowledgments

We thank Cynthia Yaschine for helpful comments on the manuscript. This work was supported by NIH grants NS28478, HD32116 and GM07524 to B.S., by the Sandler Family supporting Foundation to A.A.B. and a Milton Fund Award to F.D.

References

Allen E (1912) The cessation of mitosis in the central nervous system of the albino rat. J Comp Neurol 22:547–568

Altman J (1970) Postnatal neurogenesis and the problem of neural plasticity. In: Himwich WA (ed) Developmental neurobiology. Springfield: C.C.Thomas, pp 197–237

Altman J, Das GD (1965) Autoradiographic and histological evidence of postnatal hippocampal neurogenesis in rats. J Comp Neurol 124:319–336

Alvarez-Buylla A, Buskirk DR, Nottebohm F (1987) Monoclonal antibody reveals radial glia in adult avian brain. J Comp Neurol 264:159–170

Alvarez-Buylla A, Nottebohm F (1988) Migration of young neurons in adult avian brain. Nature 335:353–354

Alvarez-Buylla A, Kirn JR, Nottebohm F (1990a) Birth of projection neurons in adult avian brain may be related to perceptual or motor learning. Science 249:1444–1446

Alvarez-Buylla A, Theelen M, Nottebohm F (1990b) Proliferation "hot spots" in adult avian ventricular zone reveal radial cell division. Neuron 5:101–109

Alvarez-Buylla A, García-Verdugo JM, Máteo A, Merchant-Larios H (1998) Primary neural precursors and intermitotic nuclear migration in the ventricular zone of adult canaries. J Neurosci 18:1020–1037

Alvarez-Buylla A, Garcia-Verdugo JM, Tramontin, AD (2001) A unified hypothesis on the lineage of neural stem cells. Nature Rev Neurosci 2:287–293

Barami K, Iversen K, Furneaux H, Goldman SA (1995) Hu protein as an early marker of neuronal phenotypic differentiation by subependymal zone cells of the adult songbird forebrain. J Neurobiol 28:82–101

Bayer SA, Yackel JW, Puri PS (1982) Neurons in the rat dentate gyrus granular layer substantially increase during juvenile and adult life. Science 216:890–892

Blakemore WF, Jolly DR (1972) The subependymal plate and associated ependyma in the dog. An ultrastructural study. J Neurocytol 1:69–84

Cameron HA, Wooley CS, McEwen BS, Gould E (1993) Differentiation of newly born neuron and glia in the dentate gyrus of the adult rat. Neuroscience 56:337–344

Chiasson BJ, Tropepe V, Morshead CM, Van der Kooy D (1999) Adult Mammalian forebrain ependymal and subependymal cells demonstrate proliferative potential, but only sub-ependymal cells have neural stem cell characteristics. J Neurosci 19:4462–4471

Cohen E, Meininger V (1987) Ultrastructural analysis of primary cilium in the embryonic nervous tissue of mouse. Int J Dev Neurosci 5:43–51

Conover JC, Doetsch F, García-Verdugo JM, Gale NW, Yancopoulos GD, Alvarez-Buylla A (2000) Disruption of Eph/ephrin signaling affects migration and cell proliferation in the subventricular zone of the adult mouse brain. Nature Neurosci 3: 1091–1097

Craig CG, Tropepe V, Morshead CM, Reynolds BA, Weiss S, Van der Kooy D (1996) In vivo growth factor expansion of endogenous subependymal neural precursor cell populations in the adult mouse brain. J Neurosci 16:2649–2658

Doetsch F, Alvarez-Buylla A (1996) Network of tangential pathways for neuronal migration in adult mammalian brain. Proc Natl Acad Sci USA 93:14895–14900

Doetsch F, Garcia-Verdugo JM, Alvarez-Buylla A (1997) Cellular composition and three-dimensional organization of the subventricular germinal zone in the adult mammalian brain. J Neurosci 17:5046–5061

Doetsch F, Caille I, Lim DA, García-Verdugo JM, Alvarez-Buylla A (1999a) Subventricular zone astrocytes are neural stem cells in the adult mammalian brain. Cell 97:1–20

Doetsch F, Garcia-Verdugo JM, Alvarez-Buylla A (1999b) Regeneration of a germinal layer in the adult mammalian brain. Proc Natl Acad Sci USA 96:11619–11624

Doetsch F, Petreanu L, Caille I, Garcia-Verdugo JM, Alvarez-Buylla A (2002a) EGF converts transit-amplifying neurogenic precursors in the adult brain into multipotent stem cells. Neuron 36: 1021–1034

Doetsch F, Verdugo JM, Caille I, Alvarez-Buylla A, Chao MV, Casaccia-Bonnefil P (2002b) Lack of the cell-cycle inhibitor p27Kip1 results in selective increase of transit-amplifying cells for adult neurogenesis. J Neurosci 22:2255–2264

Eriksson PS, Perfilieva E, Bjork-Eriksson T, Alborn A, Nordborg C, Peterson DA, Gage FH (1998) Neurogenesis in the adult human hippocampus. Nature Med 4:1313–1317

Gage FH (2000) Mammalian neural stem cells. Science 287:1433–1438

Gage FH, Kempermann G, Palmer T, Peterson DA, Ray J (1998) Multipotent progenitor cells in the adult dentate gyrus. J Neurobiol 36:249–266

Goldman JE (1995) Lineage, migration, and fate determination of postnatal subventricular zone cells in the mammalian CNS. J.Neurooncol. 24:61–64

Goldman SA, Nottebohm F (1983) Neuronal production, migration, and differentiation in a vocal control nucleus of the adult female canary brain. Proc Natl Acad Sci USA 80:2390–2394

Gould E, Cameron HA, Daniels DC, Wooley CS, McEwen BS (1992) Adrenal hormones suppress cell division in the adult rat dentate gyrus. J Neurosci 12:3642–3650

Gould E, Cameron HA, McEwen BS (1994) Blockade of NMDA receptors increases cell death and birth in the developing rat dentate gyrus. J Comp Neurol 340:551–565

Gould E, Reeves AJ, Graziano MSA, Gross CG (1999) Neurogenesis in the neocortex of adult primates. Science 286:548–552

Gritti A, Frolichsthal-Schoeller P, Galli R, Parati EA, Cova L, Pagano SF, Bjornson CRR, Vescovi AL (1999) Epidermal and fibroblast growth factors behave as mitogenic regulators for a single multipotent stem cell-like population from the subventricular region of the adult mouse forebrain. J Neurosci 19:3287–3297

Holland EC, Varmus HE (1998) Basic fibroblast growth factor induces cell migration and proliferation after glia-specific gene transfer in mice. Proc Natl Acad Sci USA 95:1218–1223

Huang L, DeVries GJ, Bittman EL (1998) Photoperiod regulates neuronal bromodeoxyuridine labeling in the brain of a seasonally breeding mammal. J Neurobiol 36:410–420

Irvin DK, Zurcher SD, Nguyen T, Weinmaster G, Kornblum HI (2001 Expression patterns of Notch1, Notch2, and Notch3 suggest multiple functional roles for the Notch-DSL signaling system during brain development. J Comp Neurol 436: 167–181

Johansson CB, Momma S, Clarke DL, Risling M, Lendahl U, Frisén J (1999) Identification of a neural stem cell in the adult mammalian central nervous system. Cell 96:25–34

Kaplan M, Bell D (1984) Mitotic neuroblasts in the 9 day old and 11 month old rodent hippocampus. J Neurosci 4:1429–1441

Kaplan MS, Hinds JW (1977) Neurogenesis in the adult rat: electron microscopic analysis of light radioautographs. Science 197:1092–1094

Kempermann G, Kuhn HG, Gage FH (1997) More hippocampal neurons in adult mice living in an enriched environment. Nature 386:493–495

Kirn JR, Nottebohm F (1993) Direct evidence for loss and replacement of projection neurons in adult canary brain. J Neurosci 13:1654-1663

Kirschenbaum B, Doetsch F, Lois C, Alvarez-Buylla A (1999) Adult subventricular zone neuronal precursors continue to proliferate and migrate in the absence of the olfactory bulb. J.Neurosci. 19:2171–2180

Kornack DR, Rakic P (2001) The generation, migration, and differentiation of olfactory neurons in the adult primate brain. Proc Natl Acad Sci USA 98: 4752–4757

Kosaka T, Hama K (1986) Three-dimensional structure of astrocytes in the rat dentate gyrus. J Comp Neurol 249:242–260

Kuhn HG, Dickinson-Anson H, Gage FH (1996) Neurogenesis in the dentate gyrus of the adult rat: age-related decrease of neuronal progenitor proliferation. J Neurosci 16:2027–2033

Kuhn HG, Winkler J, Kempermann G, Thal LJ, Gage FH (1997) Epidermal growth factor and fibroblast growth factor-2 have different effects on neural progenitors in the adult rat brain. J.Neurosci. 17:5820–5829

Laywell ED, Rakic P, Kukekov VG, Holland EC, Steindler DA (2000) Identification of a multipotent astrocytic stem cell in the immature and adult mouse brain. Proc Natl Acad Sci USA 97: 13883–13888

Levison SW, Goldman JE (1993) Both oligodendrocytes and astrocytes develop from progenitors in the subventricular zone of postnatal rat forebrain. Neuron 10:201–212

Levitt PR, Cooper ML, Rakic P (1981) Coexistence of neuronal and glial precursor cells in the cerebral ventricular zone of the fetal monkey: an ultrastructural immunoperoxidase analysis. J Neurosci 1:27–39

Lewis PD (1968) Mitotic activity in the primate subependymal layer and the genesis of gliomas. Nature 217:974–975

Lim D A, Alvarez-Buylla A. (1999)Interaction between astrocytes and adult subventricular zone precursors stimulates neurogenesis. Proc Natl Acad Sci USA 96:7526–7531

Lim DA, Tramontin A.D, Trevejo JM, Herrera DG, García-Verdugo JM, Alvarez-Buylla A (2000) Noggin antagonizes BMP signaling to create a niche for adult neurogenesis. Neuron 28: 713–726

Lois C, Alvarez-Buylla A (1993) Proliferating subventricular zone cells in the adult mammalian forebrain can differentiate into neurons and glia. Proc Natl Acad Sci USA 90:2074–2077

Lois C, Alvarez-Buylla A (1994) Long-distance neuronal migration in the adult mammalian brain. Science 264:1145–1148

Lois C, Garcia-Verdugo JM, Alvarez-Buylla A (1996) Chain migration of neuronal precursors. Science 271:978–981

Luskin MB (1993) Restricted proliferation and migration of postnatally generated neurons derived from the forebrain subventricular zone. Neuron 11:173–189

Magavi SS, Leavitt BR, Macklis JD (2000) Induction of neurogenesis in the neocortex of adult mice. Nature 405:951–955

Malatesta P, Hartfuss E, Gotz, M (2000) Isolation of radial glial cells by fluorescent-activated cell sorting reveals a neuronal lineage. Development 127: 5253–5263

McDermott KWG, Lantos PL (1990) Cell proliferation in the subependymal layer of the postnatal marmoset, *Callithrix jacchus*. Dev Brain Res 57:269–277

McKay R (1997) Stem cells in the central nervous system. Science 276:66–71

Miyata T, Kawaguchi A, Okano H, Ogawa M (2001) Asymmetric inheritance of radial glial fibers by cortical neurons. Neuron 31: 727–741

Moreno-Lopez B, Noval JA, Gonzalez-Bonet LG, Estrada C (2000) Morphological bases for a role of nitric oxide in adult neurogenesis. Brain Res 869(1-2): 244–250

Morshead CM, Van der Kooy D (1992) Postmitotic death is the fate of constitutively proliferating cells in the subependymal layer of the adult mouse brain. J Neurosci 12:249–256

Morshead CM, Reynolds BA, Craig CG, McBurney MW, Staines WA, Morassutti D, Weiss S, Van der Kooy D (1994) Neural stem cells in the adult mammalian forebrain: a relatively quiescent subpopulation of subependymal cells. Neuron 13:1071–1082

Nait-Oumesmar B, Decker L, Lachapelle F, Avellana-Adalid V, Bachelin C, Van Evercooren AB (1999) Progenitor cells of the adult mouse subventricular zone proliferate, migrate and differentiate into oligodendrocytes after demyelination. Eur J Neurosci11:4357–4366

Noctor SC, Flint AC, Weissman TA. Dammerman RS, Kriegstein AR (2001) Neurons derived from radial glial cells establish radial units in neocortex. Nature 409:714–720

Noctor SC, Flint AC, Weissman TA, Wong WS, Clinton BK, Kriegstein AR (2002) Dividing precursor cells of the embryonic cortical ventricular zone have morphological and molecular characteristics of radial glia. J Neurosci 22:3161–3173

Nottebohm F (1985) Neuronal replacement in adulthood. In: Nottebohm F (ed) Hope for a new neurology. New York: New York Academy of Sciences, pp. 143–161

Nottebohm F, Alvarez-Buylla A (1993) Neurogenesis and neuronal replacement in adult birds. In: Cuello AC (ed) Neuronal cell death and repair. Amsterdam: Elsevier, pp. 227–236

Palmer TD, Takahashi J, Gage FH (1997) The adult rat hippocampus contains primordial neural stem cells. Mol Cell Neurosci 8:389–404

Palmer TD, Markakis EA, Willhoite AR, Safar F, Gage FH (1999) Fibroblast growth factor-2 activates a latent neurogenic program in neural stem cells from diverse regions of the adult CNS. J Neurosci 19:8487–8497

Palmer TD, Willhoite AR, Gage FH (2000) Vascular niche for adult hippocampal neurogenesis. J Comp Neurol 425:479–494

Parent JM, Yu TW, Leibowitz RT, Geschwind DH, Sloviter RS, Lowenstein DH (1997) Dentate granule cell neurogenesis is increased by seizures and contributes to aberrant network reorganization in the adult rat hippocampus. J Neurosci 17:3727–3738

Paton JA, Nottebohm F (1984) Neurons generated in the adult brain are recruited into functional circuits. Science 225:1046–1048

Privat A (1977) The ependyma and subependymal layer of the young rat: a new contribution with freeze-facture. Neuroscience 2:447–457

Privat A, Leblond CP (1972) The subependymal layer and neighboring region in the brain of the young rat. J Comp Neurol 146:277–302

Rakic P (1972) Mode of cell migration to the superficial layers of fetal monkey neocortex. J Comp Neurol 145:61–84

Rakic P (1988) Specification of cerebral cortical areas. Science 241:170–176

Ramón y Cajal S (1911) Histologie du Système Nerveux de l'Homme et des Vertébrés. Paris: Maloine

Rietze RL, Valcanis H, Brooker GF, Thomas T, Voss AK, Bartlett PF (2001) Purification of a pluripotent neural stem cell from the adult mouse brain. Nature 412:736–739

Schmechel DE, Rakic P (1979) A Golgi study of radial glia cells in developing monkey telencephalon: Morphogenesis and transformation into astrocytes. Anat Embryol 156:115–152

Seri B, Alvarez-Buylla A (2002) Neural stem cells and the regulation of neurogenesis in the adult hippocampus. Clin Neurosci Res 2:11–16

Seri B, Garcia-Verdugo JM, McEwen BS, Alvarez-Buylla A (2001) Astrocytes give rise to new neurons in the adult mammalian hippocampus. J Neurosci 21: 7153–7160

Shingo T, Gregg C, Enwere E, Fujikawa H, Hassam R, Geary C, Cross JC, Weiss S (2003) Pregnancy-stimulated neurogenesis in the adult female forebrain mediated by prolactin. Science 299: 117–120

Skogh C, Eriksson C, Kokaia M, Meijer XC, Wahlberg LU, Wictorin K, Campbell K (2001) Generation of regionally specified neurons in expanded glial cultures derived from the mouse and human lateral ganglionic eminence. Mol Cell Neurosci 17:811–820

Smart I (1961) The subependymal layer of the mouse brain and its cell production as shown by radioautography after thymidine-H3 injection. J Comp Neurol 116:325–348

Smart I, Leblond CP (1961) Evidence for division and transformations of neuroglia cells in the mouse brain, as derived from radioautography after injection of thymidine-H3. J Comp Neurol 116:349–367

Smith MT, Pencea V, Wang Z, Luskin MB, Insel TR (2001) Increased number of BrdU-labeled neurons in the rostral migratory stream of the estrous prairie vole. Horm Behav 39: 11-21

Song H, Stevens CF, Gage FH (2002) Astroglia induce neurogenesis from adult neural stem cells. Nature 417:39–44

Szele FG, Chesselet MF (1996) Cortical lesions induce an increase in cell number and PSA-NCAM expression in the subventricular zone of adult rats. J Comp Neurol 368:439–454

Tropepe V, Craig CG, Morshead CM, Van der Kooy D (1997) Transforming growth factor-α null and senescent mice show decreased neural progenitor cell proliferation in the forebrain subependyma. J Neurosci 17:7850–7859

Van Lookeren Campagne M, Gill R (1998) Tumor-suppressor p53 is expressed in proliferating and newly formed neurons of the embryonic and postnatal rat brain: comparison with expression of the cell cycle regulators $p21^{Waf1/Cip1}$, $p27^{Kip1}$, $p57^{Kip2}$, $p16^{Ink4a}$, cyclin G1, and the proto-oncogene Bax. J Comp Neurol 397:181–198

Van Praag H, Christie BR, Sejnowski, TJ, Gage FH (1999) Running enhances neurogenesis, learning, and long-term potentiation in mice. Proc Natl Acad Sci USA 96: 13427–13431

Voigt T (1989) Development of glial cells in the cerebral wall of ferrets: Direct tracing of their transformation from radial glia into astrocytes. J Comp Neurol 289:74–88

Weiss S, Dunne C, Hewson J, Wohl C, Wheatley M, Peterson AC, Reynolds BA (1996) Multipotent CNS stem cells are present in the adult mammalian spinal cord and ventricular neuroaxis. J Neurosci 16:7599–7609

Wichterle H, Garcia-Verdugo JM, Alvarez-Buylla A (1997) Direct evidence for homotypic, glia-independent neuronal migration. Neuron 18:779–791

What is the Functional Role of New Neurons in the Adult Dentate Gyrus?

G. Kempermann[1] and *L. Wiskott*[2]

Summary

The exponential growth in research results in the area of regulation of adult hippocampal neurogenesis, the life-long addition of new neurons to the hippocampal dentate gyrus, is paralleled by an increasing puzzlement about the potential function of these new cells. To determine the functional relevance of these new neurons, several fundamental problems have to be overcome. Two of them are discussed here. First, it will remain impossible to define the functional contribution the new neurons in the dentate gyrus make to hippocampal function as long as we do not know how the dentate gyrus itself contributes to hippocampal function. Our hypothesis is that adult hippocampal neurogenesis serves to avoid a stability-plasticity dilemma between learning new information and preserving old information, by allowing the dentate gyrus to adapt to new input pattern statistics while preserving the ability to process old patterns appropriately. Second, we still do not know whether, in adult neurogenesis, the structural alteration follows a specific functional stimulus and serves to consolidate a functional change triggered by that stimulus, or if less specific stimuli of novelty or complexity induce more general structural changes that prophylactically prepare the ground to better process information in similar novel or more complex situations in the future. Herein, our experimental findings and theoretical considerations argue for the latter possibility.

Discussing the rapidly growing knowledge about the regulation of neurogenesis in the adult hippocampus, a recent commentary in Science magazine was titled, "Newborn neurons search for meaning," and wondered whether, in all the "hubbub surrounding the field," the crucial fact that the function of the new neurons is still not known has not been lost (Barinaga 2003). This oversight is, of course, not the case, but the pointed statement nevertheless elucidates the state of the field.

---

[1] Max Delbrück Center for Molecular Medicine (MDC) Berlin Buch, and Volkswagen Foundation Research Group, Dept. of Experimental Neurology, Humboldt University Berlin, Germany, e-mail: gerd.kempermann@mdc-berlin.de

[2] Volkswagen Foundation Research Group, Institute for Theoretical Biology, Humboldt University Berlin, Germany, e-mail: wiskott@biologie.hu-berlin.de

Gage et al.
Stem Cells in the Nervous System:
Functional and Clinical Implications
© Springer-Verlag Berlin Heidelberg 2004

Against this criticism one could argue how it would be possible to know what the function of new neurons in the adult dentate gyrus might be if there is still no clear concept of what the dentate gyrus itself is good for. Despite years of fruitful research, no unifying theory of hippocampal function exists that would integrate all the different aspects that various experiments have brought up. However, one thing that most scientists working on the hippocampus will agree on is that the role of the dentate gyrus within hippocampal function is even more mysterious than hippocampal function altogether. The idea behind this article is not to prematurely take sides in these discussions, but to briefly review some principal problems that we are facing when we try to find out what the functions of new neurons in the adult dentate gyrus might be. After all, the attempt to understand the function of new hippocampal neurons involves entering the venerable discussion on hippocampal function by the back door.

The pessimistic assumption that adult neurogenesis is merely an atavism in a brain region that somehow missed the transition into adulthood has constantly lost ground, because adult hippocampal neurogenesis produces new granule cells in a suggestive activity-dependent and somehow function-dependent manner. Adult neurogenesis correlates with parameters describing the acquisition of the Morris water maze task, a hippocampus-dependent learning test (Kempermann and Gage 2002). This finding makes intuitive sense, given the fact that at least to some degree the hippocampus has to be considered necessary machinery for learning, for the acquisition of memories. The often-used term, "the gateway to memory" indicates just this function of the hippocampus. The hippocampus is considered to play a prime role in the formation and consolidation of declarative memory, whereas non-declarative memory does not require processing in the hippocampus. Spatial memory, which is necessary for successfully navigating the Morris water maze, is a form of declarative memory that can be assessed in rodents. Memories themselves are generally thought to be stored in the associative areas of the neocortex and not in the hippocampus, because damage to the hippocampus does not abolish the ability to recall memories stored before the damage occurred. Accordingly, in our experiments, adult neurogenesis did not show correlation with any parameters used to describe the recall of learned information in the water maze task (Kempermann and Gage 2002). These findings might be suggestive, but they cannot prove a causal relationship.

A theoretically very straightforward way of assessing the function of new neurons within the context of hippocampal function would be to simply abolish all the new cells. Hippocampal function could then be tested in animals without new neurons, by means of several classical behavioral tasks, including but not restricted to the Morris water maze. In practice it turned out that this experiment, which was published by Elizabeth Gould and her group (Shors et al., 2001), was sensitive to confounding influences (in particular, side effects of the cytostatic drug used) and reported somewhat ambiguous and conflicting results. Killing progenitor cells in the dentate gyrus with a cytostatic drug did not affect learning the water maze (Shors et al. 2002) but did affect a hippocampus-dependent ("trace") form of eye-blink conditioning while sparing the hippocampus-independent ("delayed") form (Shors et al.2001). It is not easy to understand why neurogenesis would be necessary to mediate a rather simple conditioned reflex that probably requires minimal processing, to say nothing of structural reorganization in the hippocampus, but not be necessary for a complex, hippocampal learning task with a long acquisition phase and the known ability to influence hippocampal morphology, including an induction of adult neurogenesis. Therefore, at present the most stringent conclusion from these experiments might be that more experiments are needed. One fundamental problem lies in the fact that the experimental procedure in these two published studies did not eliminate the new neurons but rather the proliferating (progenitor) cells of the dentate gyrus – and elsewhere. If one were interested in studying the role of the rooster in the chicken farm, smashing eggs might not be the way to go.

The newly generated neurons of the adult dentate gyrus are granule cells, but do they differ from older granule cells? Are there functionally two types of granule cells: one larger population that is generated during development and early postnatally, and one generated during adulthood? Obviously, an activity- and function-dependent production of new neurons can only take place when the individual is capable of activity. This characteristic alone sets adult neurogenesis apart from embryonic neurogenesis, where the amount of activity is nil, limited or at least qualitatively different. Also, the population of progenitor cells, from which the new cells are derived in adult neurogenesis, appears to be distinct. While during embryogenesis the dentate gyrus is formed from the side that later becomes the molecular layer, adult neurogenesis adds new neurons in the subgranular zone. These factors notwithstanding, by all standards the new granule cells appear to become indistinguishable from older granule cells. They express the same mature markers, in particular calbindin (Kuhn et al.,1996), they extend their axons along the mossy fiber tract (Stanfield and Trice 1988; Hastings and Gould 1999; Markakis and Gage,1999), and they have electrophysiological characteristics that are very similar to those of older granule cells (van Praag et al. 2002). Particularly with regard to the last criterion, however, the matter is complicated. In their landmark study, van Praag and coworkers labeled new neurons with a retrovirus carrying the green-fluorescent protein (GFP) and used the patchclamp technique to analyze the new cells and the neighboring older granule cells (van Praag et al. 2002). However, with this approach the population of new cells could not be studied exhaustively. Therefore the question of whether the new granule cells functionally mirror the range defined by the granule cell layer in general remained unanswered. Do the new cells reflect the local differences found in the granule cell layer (Wang et al. 2000)? Input to the granule cell layer varies considerably, and granule cells receive afferents with different neurotransmitter systems (glutamatergic, dopaminergic, acetylcholinergic, and serotonergic) in a varying combination. The main (glutamatergic) input from the entorhinal cortex via the perforant path is not homogeneous either. Within the thickness of the granule cell layer, the electrophysiological properties of the granule cells change (Wang et al. 2000), potentially indicating that granule cells differ to some degree. New granule cells can be found in the entire granule cell layer, with only minimal local preferences (slightly more new cells in the dorsal, suprapyramidal blade than in the ventral, infrapyramidal blade; Kempermann et al. 2003). Moreover, it seems that within days or weeks into their development the new cells find their place within the granule cell layer (Kempermann et al. 2003). After that their distribution does not change over time. This finding could imply that the new granule cells acquire a locally specific functional phenotype. The alternative hypothesis is that the new cells are in a class of their own, with subtle differences from all other granule cells, and are perhaps independent of their individual location. The answer to these questions will clearly influence the final interpretation of what adult neurogenesis is good for.

Adult hippocampal neurogenesis does not contribute vast numbers of new neurons. But there is increasing evidence that the granule cell layer grows measurably during those periods postnatally when adult neurogenesis occurs at a high rate. As adult neurogenesis declines with age, to very low levels in old age, the contribution becomes unmeasurable (Kempermann et al. 1998). A rough estimate is that the total growth of the dentate gyrus in adult mice is about 10 to 20%, reflecting something like 30,000 to 60,000 neurons. These are too few cells to allow the build-up of entire new structures. There is no evidence, for example, that the adult dentate gyrus would be expanded in certain areas, in a continuation of late embryonic and early postnatal development. Most importantly, there is no indication of a turnover in the dentate gyrus (Biebl et al. 2000; Cooper-Kuhn et al. 2002).

Also, the new cells appear to be rather evenly distributed throughout the rostro-caudal extension of the granule cell layer. However, a detailed analysis has not been conducted so far, so there still might be distributions that are not immediately obvious. The distribution within the thickness of the granule cell layer, however, has been studied. The majority of new granule cells (50–60%) is found in the inner third of the granule cell layer and remains there (Kempermann et al. 2003). These findings suggest that function might be dependent on location (rather than age), but more detailed experiments are needed to confirm this idea. However, from such observations we have derived the hypothesis that the new neurons are unlikely to add bulk memory or multiply processing power but are strategically inserted into the neuronal network of the granule cell layer (Kempermann 2002). The mossy fiber tract, which connects the dentate gyrus to area CA3, is one of the bottlenecks within the information processing circuits of the hippocampus. Accordingly, the relative contribution of single new cells to overall function could be greatest at exactly this location.

Information flows from cortical regions via the entorhinal cortex to the dentate gyrus, and from there to CA3, onward through the Shaffer collaterals to CA1 (and the subiculum) and from there via the entorhinal cortex back out to the associative areas of the cortex (Amaral and Witter1989). Our theory is that adult neurogenesis allows the hippocampus to adapt the first of its three processing modules to the level of complexity and novelty frequently encountered by the animal (Kempermann 2002). Thus adult neurogenesis would modify the network by inserting relatively few new neurons rather than primarily increasing its size. Increasing processing power of a computer by 10% would be ineffective. The main effect of adult neurogenesis will be qualitative and not quantitative, and it will be cumulative improving the quality of the network with increasing age. Old animals that have very low levels of adult neurogenesis would not need as substantial adaptations as younger ones. Our studies have shown, however, that in old animals neurogenesis can be stimulated to a much larger relative degree than in younger animals (Kempermann et al. 1998, 2002), suggesting that upon functional challenge the hippocampus recruits the maximum number of new cells available to cope with the new needs. Feng et al. (2001) have proposed that adult neurogenesis provides a means of erasing old memories and preparing the dentate gyrus for new information. However, the findings that there is no obvious turnover of cells and that the increase in neuron numbers is incremental, slow and relatively modest in absolute terms argue against this idea. Also, old animals learn hippocampal tasks better than the low level of adult neurogenesis at old age would suggest.

There is no indication that the dentate gyrus stores memories. This role is usually assigned to CA3, which has strong recurrent connectivity. CA3 is commonly conceptualized as a so-called Hopfield network (Treves and Rolls 1994), a type of recurrent, artificial neuronal network that can store patterns quickly and recall them in an auto-associative fashion, meaning that even a partial cue can retrieve a complete pattern. Hopfield networks work particularly efficiently if the patterns to be stored are uncorrelated or orthogonal (Hertz et al. 1991). Thus, orthogonalizing the input patterns could be an important function of the dentate gyrus to increase storage capacity of the CA3 network and, more importantly, to reduce interference between different patterns. One can think of this orthogonalization as a transformation that removes common features and emphasizes differences to make the transformed patterns maximally unrelated. This transformation can be achieved only in a statistical sense, of course, and not for each pattern individually. This hypothesis is important because it helps us understand why a quantity of new neurons is not needed for each bit of new information processed.

One way of orthogonalizing patterns is to make their representations sparse (Hertz et al. 1991); in fact, the dentate gyrus is known to have a very sparse neural activity (Barnes et al. 1990). Maximizing sparseness is closely related to independent component analysis (ICA), which attempts to decompose patterns linearly into statistically independent components. Sparseness and ICA have been suggested as important coding principles in the primary visual (Bell and Sejnowski 1997; Olshausen and Field 1997) and auditory cortex (Lewicki 2002), and ICA has been demonstrated to also be advantageous for pattern discrimination (Bartlett et al. 2002). Thus we follow the hypothesis that orthogonalization is achieved by sparsification (Barnes et al. 1990; Treves and Rolls 1994) and assume that this is related to an implicit ICA.

But why might neurogenesis be required for this function? The transformation necessary to orthogonalize input patterns depends on the statistics of the patterns. Assume the dentate gyrus of a mouse has developed an orthogonalizing transformation for a given range of input patterns let's say for a given environment the mouse lives in. Then the mouse is transferred into a qualitatively different environment. The dentate gyrus now has to orthogonalize a different range of input patterns and should learn a new, adapted transformation. However, since the transformation is not only necessary for storage but also for recall, the mouse also needs the old transformation; otherwise it could not recall memories from the old environment. This is a classical stability-plasticity dilemma. Part of our hypothesis is that neurogenesis is one way of solving this dilemma. The idea is that, instead of adapting all synaptic weights in the dentate gyrus network, which would erase the old transformation, the old synaptic weights are kept fixed and new neurons with new synaptic connections are added. This hypothesis is somewhat in the spirit of cascade-correlation learning architecture (Fahlmann and Lebiere 1990). The new neurons would literally add new (orthogonal) dimensions to the representation that have not been present or important in the old environment. This hypothesis also explains why neurogenesis may decrease with increasing age, because the effect of adding new dimensions to the representation accumulates and tends to saturate. Notice that our hypothesis that neurogenesis serves learning of new transformations while preserving old transformations does not depend on whether CA3 is considered a permanent storage or an intermediate one (see below), as long as the time of storage is longer than the time scale on which the animal has to adapt to qualitatively new environments.

However, there is a fundamental problem in the hypothesis that the new neurons participate in orthogonalization or any other similar processing step, if the input that is to be processed is the one that has to trigger neurogenesis to achieve successful processing. It takes weeks or perhaps months for a new neuron to become fully integrated and functional (van Praag et al. 2002). The input that will benefit from the increased number of neurons cannot be the same that once triggered neurogenesis and thus provided the new neurons whose synaptic weights can now be altered. Thus, the situation is somewhat similar to data from research on long-term-potentiation (LTP), the putative electrophysiological mechanism underlying learning. There, measurable synaptic changes precede detectable structural changes, that is the formation of new dendritic spines (Engert and Bonhoeffer 1999). Accordingly, there are two general options. Either adult neurogenesis is involved in consolidating an earlier functional change on a structural level, or adult neurogenesis is a non-specific preparatory step that allows the general adaptation of the system to experienced levels of complexity. Either structure follows function or function follows structure. Our hypothesis that old synaptic weights are fixed argues for the latter explanation, which would also explain why such unspecific stimuli such as physical activity (van Praag et al. 1999) can trigger adult neurogenesis.

If adult neurogenesis is so crucial for the dentate gyrus, why is it not required in CA1 as well, since CA1 also has to adapt to the new pattern statistics? In fact, if CA1 actually inverted the transformation carried out in the dentate gyrus, it would have to closely parallel the adaptation of the dentate gyrus. However, while the dentate gyrus has to take measures to also preserve the old transformation, CA1 does not have this constraint. CA1 simply has to follow the changes with the objective that it inverts the transformation performed by the dentate gyrus. This is similar to a so-called auto-encoder network (Hertz et al. 1991) that is, for example, trained through back-propagation and where the encoding stage corresponded to the dentate gyrus and the decoding stage to CA1 – with the difference that the encoding stage had the additional objective of orthogonalizing the input patterns.

Retrograde amnesia found in patients with bilateral hippocampal damage has been interpreted as reflecting the period the hippocampus needs to consolidate the memory. However, what makes things considerably more complicated is the finding that, whereas memories might still be retrievable in subjects with hippocampal damage, this recalled information appears to be qualitatively altered. Also, for some pieces of information, the retrograde amnesia seems to extend much further into the past than for others. This finding could indicate a different speed of consolidation for different memories, but it might also argue in favor of the idea that, in one sense or another, the hippocampus is involved in the retrieval of the already stored information or that it is the actual long-term storage site for some kinds of memories (Nadel and Moscovitch 1997). It has also been suggested that stored memory contents might again become hippocampus-dependent during retrieval and thus vulnerable to alterations (Nadel and Land 2000; Myers and Davis 2002). Additionally, the hippocampus is profoundly involved in the limbic system and presumably through these connections contributes to what we call "emotional memory" and to the affective contexts and tags we assign to new information.

In this brief discussion we have touched on two of the main problems in recognizing the function of new neurons in the adult dentate gyrus. Both are fundamental issues. First, we have discussed the function of adult neurogenesis in the context of the function of the dentate gyrus. Under the assumption that the dentate gyrus performs a transformation of entorhinal input patterns to facilitate storage in CA3, we have hypothesized that adult neurogenesis allows the dentate gyrus to adapt to new input pattern statistics while preserving the ability to process old patterns appropriately. This hypothesis provides a solution to the stability-plasticity dilemma between learning new information and preserving old information. A second question is whether adult neurogenesis acts post hoc to provide a structural consolidation of a specific, functionally induced change or prophylactically reacts to the more general experience of increased functional challenge, thereby allowing a long-term adaptation of the system to better cope with similar situations in the future. Our hypothesis that old synaptic weights should be kept constant would argue for the latter. Overall, we believe that the research on adult neurogenesis in the dentate gyrus is not only interesting in itself but also provides a new avenue for understanding hippocampal function in general.

References

Amaral DG, Witter MP (1989) The three-dimensional organization of the hippocampal formation: a review of anatomical data. Neuroscience 31:571–591

Barinaga M (2003) Developmental biology. Newborn neurons search for meaning. Science 299: 32–34

Barnes CA, McNaughton BL, Mizumori SJ, Leonard BW, Lin LH (1990) Comparison of spatial and temporal characteristics of neuronal activity in sequential stages of hippocampal processing. Prog Brain Res 83:287–300

Bartlett MS, Movellan JR, Sejnowski TJ (2002) Face recognition by independent component analyis. IEEE Trans Neur Netw 13:1450–1464

Bell AJ, Sejnowski TJ (1997) The "independent components" of natural scenes are edge filters. Vision Res 37:3327–3338

Biebl M, Cooper CM, Winkler J, Kuhn HG (2000) Analysis of neurogenesis and programmed cell death reveals a self-renewing capacity in the adult rat brain. Neurosci Lett 291:17–20

Cooper-Kuhn CM, Vroemen M, Brown J, Ye H, Thompson MA, Winkler J, Kuhn HG (2002) Impaired adult neurogenesis in mice lacking the transcription factor E2F1. Mol Cell Neurosci 21:312–323

Engert F, Bonhoeffer T (1999) Dendritic spine changes associated with hippocampal long-term synaptic plasticity. Nature 399:66–70

Fahlmann SE, Lebiere C (1990) The cascade-correlation learning architecture. In: Touretzky D (ed) Advances in neural information processing systems 2 (NIPS 1989). Morgan-Kaufmann, San Fransisco, pp 524–532

Feng R, Rampon C, Tang YP, Shrom D, Jin J, Kyin M, Sopher B, Martin GM, Kim SH, Langdon RB, Sisodia SS, Tsien JZ (2001) Deficient neurogenesis in forebrain-specific presenilin-1 knockout mice is associated with reduced clearance of hippocampal memory traces. Neuron 32:911–926

Hastings NB, Gould E (1999) Rapid extension of axons into the CA3 region by adult-generated granule cells. J Comp Neurol 413:146–154

Hertz J, Krogh A, Palmer RG (1991) Introduction to the theory of neural computation. Redwood City: Addison-Wesley.

Kempermann G (2002) Why new neurons? Possible functions for adult hippocampal neurogenesis. J Neurosci 22:635–638

Kempermann G, Gage FH (2002) Genetic determinants of adult hippocampal neurogenesis correlate with acquisition, but not probe trial performance in the water maze task. Eur J Neurosci 16:129–136

Kempermann G, Kuhn HG, Gage FH (1998) Experience-induced neurogenesis in the senescent dentate gyrus. J Neurosci 18:3206–3212

Kempermann G, Gast D, Gage FH (2002) Neuroplasticity in old age: sustained fivefold induction of hippocampal neurogenesis by long-term environmental enrichment. Ann Neurol 52:135–143

Kempermann G, Gast D, Kronenberg G, Yamaguchi M, Gage FH (2003) Early determination and long-term persistence of adult-generated new neurons in the dentate gyrus of mice. Development 130:391–399

Kuhn HG, Dickinson-Anson H, Gage FH (1996) Neurogenesis in the dentate gyrus of the adult rat: age-related decrease of neuronal progenitor proliferation. J Neurosci 16:2027–2033

Lewicki MS (2002) Efficient coding of natural sounds. Nature Neurosci 5:356–363

Markakis E, Gage FH (1999) Adult-generated neurons in the dentate gyrus send axonal projections to the field CA3 and are surrounded by synaptic vesicles. J Comp Neurol 406:449–460

Myers KM, Davis M (2002) Systems-level reconsolidation: reengagement of the hippocampus with memory reactivation. Neuron 36:340–343

Nadel L, Moscovitch M (1997) Memory consolidation, retrograde amnesia and the hippocampal complex. Curr Opin Neurobiol 7:217–227

Nadel L, Land C (2000) Memory traces revisited. Nature Rev Neurosci 1:209–212

Olshausen BA, Field DJ (1997) Sparse coding with an overcomplete basis set: a strategy employed by V1? Vision Res 37:3311–3325

Shors TJ, Miesegaes G, Beylin A, Zhao M, Rydel T, Gould E (2001) Neurogenesis in the adult is involved in the formation of trace memories. Nature 410:372–376

Shors TJ, Townsend DA, Zhao M, Kozorovitskiy Y, Gould E (2002) Neurogenesis may relate to some but not all types of hippocampal-dependent learning. Hippocampus 12:578–584

Stanfield BB, Trice JE (1988) Evidence that granule cells generated in the dentate gyrus of adult rats extend axonal projections. Exp Brain Res 72:399–406

Treves A, Rolls ET (1994) Computational analysis of the role of the hippocampus in memory. Hippocampus 4:374–391

van Praag H, Kempermann G, Gage FH (1999) Running increases cell proliferation and neurogenesis in the adult mouse dentate gyrus. Nature Neurosci 2:266–270

van Praag H, Schinder AF, Christie BR, Toni N, Palmer TD, Gage FH (2002) Functional neurogenesis in the adult hippocampus. Nature 415:1030–1034

Wang S, Scott BW, Wojtowicz JM (2000) Heterogeneous properties of dentate granule neurons in the adult rat. J Neurobiol 42:248–257

Do Forebrain Neural Stem Cells Have a Role in Mammalian Olfactory Behavior?

E. Enwere[1] and *S. Weiss*[1]

Summary

In this review/commentary, we discuss rostral forebrain neural stem cells, both with respect to their presence in the developing central nervous system and the putative role that they play in adult olfactory behavior. In cell culture and in vivo, the neural stem cells respond to fibroblast growth factor 2 (FGF-2) or epidermal growth factor (EGF) with an increase in their proliferation and self renewal. On removal of these factors, the neural stem cell progeny differentiate to form neurons, astrocytes and oligodendrocytes. In the adult, the neural stem cells reside in the subventricular zone within the walls of the lateral ventricles, where they produce neuronal precursors that migrate through the forebrain to become interneurons in the olfactory bulb (OB). We discuss the components of the olfactory system, from the olfactory epithelium, which is innervated as part of the peripheral nervous system, to the OBs in the central nervous system, as well as the connections between both. We then explore the possible functions neural stem cell-derived OB interneurons may have in behavior, in relation to olfactory discrimination and memory. Newborn olfactory interneurons have been found to be more susceptible to cell death than more mature neurons. This finding and other evidence lead us to suggest that new olfactory neurons have different functions from their older counterparts. We examine this possibility and pose further questions relevant to a clear understanding of neural stem cells and olfaction.

Neural Stem Cells

What is a Neural Stem Cell?

There is still no universally accepted definition of what constitutes a stem cell. However, the use of terms such as "pluripotency" and "self-renewal" in association with stem cells indicates that these cells are best defined by

---

[1] Genes & Development Research Group, University of Calgary Faculty of Medicine, Calgary, Alberta, Canada T2N4N1

Gage et al.
Stem Cells in the Nervous System:
Functional and Clinical Implications
© Springer-Verlag Berlin Heidelberg 2004

what they do, rather than by what they are. This mode of definition stems partly from the (albeit changing) fact that, in various instances, such as in the haematopoietic and central nervous systems, precise identification of the stem cells is problematic (Dzierzak et al. 1998; Price 2001). Nevertheless, it may be argued that the functions of a stem cell remain more important than its identity (van der Kooy and Weiss 2000). On that basis, we could tender an operational definition (one that is oversimplified and incomplete, but suitable for our purposes) that stem cells are "cells resident in an organism that retain the capacity for self-renewal, and which directly or indirectly produce progeny that reconstitute the various cell types of the host tissue" (Hall and Watt 1989; Morrison et al. 1997; Marshak et al. 2001). Adding to this definition, one may ascribe such other features to stem cells as the potential for asymmetric division (to produce another stem cell and a cell of more restricted fate), maintenance of a relatively undifferentiated phenotype (as compared to the progeny), and mitotic quiescence when proliferation is not required (Lajtha 1979; Potten and Loeffler 1990; Morrison et al. 1997). We can say that stem cells function in creation and regeneration: in creation of tissue and organs de novo, as with embryonic stem cells and the totipotent stem cell of the blastocyst (Weissman 2000; Hadjantonakis and Papaioannou 2001), and in regeneration of extant tissue, as with tissue- or organ-specific stem cells of the gut, skin, blood and olfactory epithelium (Monti-Graziadei and Graziadei 1979; Potten and Loeffler 1990).

The dogma that the mammalian central nervous system lacked the ability to regenerate and thus, by implication, lacked stem cells, held until recently when Reynolds and Weiss (1992) discovered a population of neural stem cells in the adult central nervous system. These cells reside in a remnant of the embryonic germinal zone generally known as the subventricular zone (SVZ). They self-renew indefinitely in defined media containing epidermal growth factor (EGF) and/or basic fibroblast growth factor (bFGF, or FGF2), producing cellular aggregates known as neurospheres. These spheres express the intermediate filament protein nestin but do not express antigenic markers of differentiated cellular phenotypes. However, on removal of mitogens from the growth medium, the cells differentiate into neurons and glia, thus meeting the rough characteristic criteria of stem cells established above. When transplanted into the cerebrum, neural stem cells also differentiate into neurons and glia, though the glial fate appears to be preferred (Hammang et al. 1997; Winkler et al. 1998). While neural stem cells proliferate extensively in culture media, they are quiescent in vivo, dividing very rarely – unless stimulated. They exist in the SVZ with their progeny, which are a population of rapidly proliferating, but lineage-restricted, progenitor cells. When these progenitors are destroyed with tritiated thymidine or cytosine arabinofuranoside, the stem cells remain unaffected but become actively proliferative to regenerate the germinal zone within 8 to 10 days (Morshead et al. 1994; Doetsch et al. 1999). Neural stem cell proliferation in the SVZ can also be stimulated with intracerebroventricular infusions of EGF (Craig et al. 1996).

Neural Stem Cells in Development

As one would expect during development, neural stem cells of various sorts abound in the embryonic CNS. Cattaneo and McKay (1990) isolated neuronal stem cells that proliferated in response to FGF2 and, on FGF withdrawal, differentiated into neurons. Other groups have isolated multipotent stem cells from E12 and E14 rat cerebral cortices (Davis and Temple 1994), from the developing spinal cord (Represa et al. 2001) and from E14 mouse striatal primordia (Reynolds et al. 1992; Reynolds and Weiss 1996).

Neural stem cells emerge remarkably early in CNS development, as early as E8.5 in the mouse anterior neural plate (Tropepe et al. 1999), a stage at which expression of the FGF receptor (FGFR1) largely localizes to the ventricular region of the forming neural tube (Wanaka et al. 1991). This stem cell could only be isolated in FGF2, but passaged into EGF, suggesting that EGF-responsive stem cells arose from a more primitive population of FGF-responsive cells. Thus, EGF- and FGF-responsive stem cells are present in the germinal zone by E14, making up part of a heterogeneous population of proliferating cells (Reynolds et al. 1992; Vescovi et al. 1993). However, the eventual responsiveness to EGF has led some to suggest that there are two stem cell populations at this stage: one that proliferates in response to FGF, through the mediation of the FGFR1, and another that responds to EGF through the EGF receptor (EGFR). Some evidence in support of this hypothesis includes work that shows that, at low cell densities, FGF and EGF have an additive effect on stem cell proliferation, in that more stem cells proliferate in the presence of both growth factors than with either alone (Tropepe et al. 1999). Further, Kilpatrick and Bartlett (1995) isolated a precursor that was multipotent in response to FGF but glial-restricted with EGF. Others suggest that FGF merely accelerates development of EGF-responsive stem cells (Lillien and Raphael 2000). The exact circumstances surrounding EGF and FGF responsiveness, as with many such issues in the neural stem cell field, remain shrouded in controversy.

While the precise identity of the embryonic neural stem cell remains elusive, its function is presumably more straightforward – formation of the CNS. All CNS cells arise from the neuroepithelium, and largely from neuroepithelial cells, which are the primary cell types in the developing neural tube (Huttner and Brand 1997). These cells characteristically retain portions of their plasma membranes at the inner (ventricular) and outer (pial) surfaces of the neural tube. They may divide symmetrically early in CNS development to generate sufficient cell numbers (McConnell 1995; Rakic 1995), whereupon, after closure of the neural tube, they produce radial glia, neurons, and various progenitor cells resident in the embryonic striatal ventricular zone. While, as reported earlier, neural stem cells exist in the neural tube as early as E8.5 (Tropepe et al. 1999), the extent of their contribution to the proliferation seen in the neural tube, and to the eventually formation of the CNS, is unknown. However, stem cells are abundant in E8.5 spinal neural tube (Kalyani et al. 1998) and E10 telencephalon (Temple 2001), which suggests that neural stem cells contribute to both the spinal cord and basal forebrain. Indeed, Temple's group recently reported that cells in the lateral ganglionic eminence, which contributes GABAergic neurons to the cortex, are multipotent stem cells (He et al. 2001). The conclusions of this study were somewhat speculative, however, as they could not directly show that the GABAergic neurons in question arose from stem cells. As we still lack adequate methods for identifying neural stem cells in vivo, studies of this kind are bound to involve certain leaps of faith in approaches taken and in conclusions eventually drawn.

Neural Stem Cells in the Adult

Postnatally, the functions of neural stem cells change radically, and they become more spatially restricted and functionally specific. Neural stem cells in the adult CNS are largely confined to two regions: the SVZ, as previously described, and the subgranular zone in the dentate gyrus of the hippocampus (Palmer et al. 1997). We focus here on the neural stem cells of the adult forebrain SVZ. In the first few weeks of life, there are two functional regions of the SVZ, based on the phenotype of progenitors produced by resident stem cells. (This heterogeneity is lost later in life; [Doetsch and Alvarez-Buylla 1996].) The caudal region of the SVZ is primarily gliogenic, producing glial progenitors that migrate dorsally and laterally into the corpus callosum. In contrast, the SVZ towards the anterior horn of the lateral ventricle, a region designated as the SVZa, produces exclusively neuronal progenitors (Luskin 1993). These progenitors are phenotypically neurons, expressing neuronal markers such as TuJ1, but retain the ability to divide (Luskin et al. 1997). They also migrate great distances along a highly restricted pathway through the forebrain known as the rostral migratory stream (RMS), utilizing a unique mode of migration known as chain migration (Garcia-Verdugo et al. 1998). Here, the cells "slide" along each other without axonal or radial guides, at speeds of up to 120 μm per hour, traversing a 5 mm distance in a matter of days. The mechanism behind this mode of migration is unknown; however, mice lacking the polysialic acid-conjugated neural cell adhesion molecule (PSA-NCAM) experience greatly impaired migration of progenitors and accumulation of these progenitors in the forebrain (Tomasiewicz et al. 1993; Hu et al. 1996).

The destination of cells in the RMS is ultimately the olfactory bulb (OB), which is the first CNS component of the olfactory system (Lois et al. 1996). They enter through the ependymal zone of the bulb, migrate radially outwards, and differentiate into granule and periglomerular interneurons, which are the two primary interneuronal cell types of the OB. Most of these neurons are produced shortly after birth: between 3 and 6 weeks of age, the size of the OB almost doubles with the influx of new neurons. As many of these neurons undergo constitutive cell death (Kaplan et al. 1985; Kato et al. 2000), olfactory neurogenesis progresses throughout life to maintain their numbers in the bulb. Thus, at adulthood, the size of the murine OB increases only slightly with time, as the influx of new neurons from the SVZ matches the turnover rate of old neurons.

Neural Stem Cells and Olfaction

The Sense of Smell: Systems and Processes

Few individuals (gourmet chefs possibly excluded) would consider the sense of smell to be of critical importance to humans, but in animals, the ability to detect chemicals in the surrounding environment – for purposes of feeding, reproduction and social interaction – is necessary for survival itself. Included in the repertoire of substances perceivable in this fashion are innumerable largely organic molecules of variable molecular mass, including aldehydes, ketones, acids, amines and halides (Firestein 2001). This range of chemicals represents identifiers of sources of food, potential mates, danger, safety and other such rudiments of survival, as well as the learning of, and memory for, the same. Not surprisingly, therefore, there is a remarkable degree of sophistication and detail nested in the olfactory system of even the most basic eukaryotes.

Given the evolutionary importance of the sense of smell, it should be no surprise that there is a significant measure of evolutionary convergence in olfaction among eukaryotes. In most cases, nonetheless, the process is a variant on this: odor molecules dissolve in the aqueous medium of the olfactory epithelium and bind to an assortment of G-protein-coupled receptors that activate a cyclic AMP second-messenger pathway. The olfactory receptor (OR) proteins themselves were identified 13 years ago (Buck and Axel 1991) and were found to constitute a massive family of about 1,000 genes in the mammalian genome, though in humans about two-thirds of these are pseudogenes (Firestein 2001). The process of signal transduction in the olfactory epithelium is poorly understood; nevertheless, odor information is carried from the olfactory epithelium by the olfactory receptor neuron (ORN), which is a bipolar neuron in the peripheral nervous system that extends an axon to the CNS (Steinbrecht 1969). The basic ORN structure in vertebrates is almost identical to that in arthropods, nematodes and even insects (Hildebrand and Shepherd 1997), in all of which the processes of olfaction bear strong similarities to those of vertebrates.

A key to an understanding of how the olfactory system works lies in the possibility that a topographic map of odor molecules exists that is established and maintained throughout the olfactory pathway. Recent studies have established that there is a measure of topographic organization in the olfactory epithelium and OB. ORNs send their axons to the OB or, specifically, to circular collections of neuropil known as glomeruli. Each ORN expresses only one of the ORs and, remarkably, ORNs expressing the same OR converge on one or two specific glomeruli (Ressler et al. 1994; Mombaerts et al. 1996), occasionally taking unusually long and circuitous routes to the target (Treloar et al. 2002) even after the neurons are genetically ablated and restored (Gogos et al. 2000). This process implies that each odor molecule activates a restricted set of glomeruli as part of an olfactory code established within the olfactory bulb itself. This restriction may provide the initial basis of olfactory discrimination, or various other olfactory processes mediated in the bulb.

The ORNs synapse on mitral and tufted cells, which are the second-order excitatory projection neurons of the OB (Shipley and Ennis 1996). These neurons in turn interact with two populations of inhibitory interneurons, the granule and periglomerular neurons. The dendrites of mitral cells make synaptic connections with the dendrites of granule cells within the external plexiform layer (Jackowski et al. 1978; Price and Powell 1970). The granule-to-mitral part of the dendrodendritic synapse is inhibitory on the secondary dendrites of the mitral cells (Jahr and Nicoll 1982). The periglomerular cells dwell within the glomeruli in the periphery of the OB, whereas the granule cells lie medially in a layer known as the granule cell layer (GCL). Both cell types express a wide variety of neurotransmitters, notably GABA (in periglomerular and granule neurons) and dopamine (in periglomerular neurons); (Shepherd and Greer 1990). Granule cells occur in tight clusters and are coupled by gap junctions (Reyher et al. 1991). These clusters are absent in neonates but form by six weeks of age. Golgi studies show that granule cells lack axons (Cajal 1911; Price and Powell 1970; Schneider and Macrides 1978) but have basal dendrites that project into the external plexiform layer of the OB. The spike activity in granule cells is similar to that of retinal amacrine cells (Mori 1987). Though granule cells exist in far greater numbers than their dopaminergic counterparts, they may have similar functions, in that both exhibit GABAergic reciprocal synapses back on mitral/ tufted cell dendrites (Schoppa et al. 1998).

Mitral cells exhibit complex responses to odor stimulation, consisting of both excitation and inhibition in the course of an odor presentation. Intracellular recordings from mitral cells show a characteristic hyperpolarization that counteracts the spontaneously occurring spikes produced by the cell (Scott 1991). Responses also vary with odor concentration (Wellis et al. 1989) and certain cells seem to respond better to particular concentrations of any odor (Kauer 1974). The concept of lateral inhibition between mitral cells is supported by evidence that cells separated by large distances within the bulb showed generally opposite responses to odor stimulation (Meredith 1986). Conversely, those in mutual proximity exhibited similar responses (Buonoviso and Chaput 1990). This may result from the inhibition exacted on the mitral cells by the interneurons. The mitral and tufted cells project directly to the primary olfactory cortex, which spawns extensive reciprocal connections to the mediodorsal nucleus of the thalamus, posterolateral orbitofrontal region, and back to the OB itself (Doty 2001).

Neurogenesis and Olfactory Discrimination

To demonstrate a function for neural stem cells in olfactory behavior, one must specifically investigate the functions of the interneurons produced by said stem cells. We can hypothesize several possible functions for these neurons. One possibility is that olfactory interneurons enhance the animal's ability to discriminate between similar odors, by refining input from the olfactory epithelium. The convergence ratio from the olfactory epithelium to the glomeruli is significant – as much as 25,000:1 in rabbits (Hildebrand and Shepherd 1997) – suggesting that refinement of the incoming signal is necessary. Early electrophysiological work by Leveteau and MacLeod (1966) demonstrated the presence of strong activation signals in some glomeruli in response to a particular stimulus and none whatsoever in others, leading them to suggest an element of olfactory discrimination in the bulb. Further studies demonstrate that mitral cells excited by a particular odor in one glomerulus may use olfactory interneurons to inhibit firing of adjacent mitral cells, a process tantamount to lateral inhibition (Meredith 1986; Wilson and Leon 1987). Studies in non-mammalian organisms also suggest such an action: in honeybees, blockade of GABAergic signalling impairs olfactory discrimination between similar odors (Hosler et al. 2000), suggesting that GABA-producing interneurons are important in odor discrimination.

Research examining the potential of olfactory interneurons in olfactory discrimination often depicts the sheer number of neurons as being directly proportional to their capabilities in behavior. Gheusi and others (2000) postulated that the NCAM mutant mouse, which, as a result of migratory deficits in the neuronal precursors, has fewer interneurons in the OB than do wild-type littermates, is unable to perform a simple olfactory discrimination task. It is possible, however, that the synaptic density, rather than the number of neurons present, mediates such a behavioral outcome. This possibility receives backing from studies carried out on the Drosophila mutant gigas (gig), which, as a result of extra rounds of DNA replication that occur after mitosis, have larger glomeruli and more associated synapses than do controls (Acebes and Ferrús 2001). Mutant flies are more sensitive to low concentrations of odorant than wild types, suggesting that the synapse density-olfactory behavior correlation holds true. This concept, however, holds the assumption that all neurons in the relevant olfactory structure are equal – an assumption, as will be addressed in a later section, that may not necessarily hold true.

Neurogenesis and Olfactory Learning

Another prospect for the function of olfactory interneurons resides in the fact that in mice, olfactory neurogenesis is most pronounced in the first two months of life (Luskin 1993). The size of the mouse OB doubles between three and six weeks of age, whereupon the RMS and SVZa undergo a significant reduction in thickness (Pencea et al. 2001). After the initial surge of OB growth, the size increases only slightly with time, and essentially attains a "steady state" in which the influx of new neurons from the SVZ matches the turnover rate of old neurons. This finding suggests that any behavioral function related to these neurons should change between birth and adulthood and thereafter would remain relatively constant. One potential candidate for such a function is olfactory memory. The early stages of life of neonates are the most important for acquisition and learning of important odors, such as the odors of the mother, the home cage, and food (Porter and Etscorn 1975; Leon 1992). This critical olfactory imprinting becomes less necessary with time, possibly resulting in a diminishment of the acuity of olfactory memory formation over a time course that may correlate with the progress of the most prolific stages of olfactory neurogenesis.

Research has long supported a connection between olfaction and cognitive function in rodents. One of the earlier links was the discovery that cells in the olfactory cortex project to the segment of the thalamic mediodorsal nucleus that connects to the orbital prefrontal cortex (Powell et al. 1965). Other studies demonstrated bulbar projections to the amygdala, entorhinal cortex and hypothalamus (Price and Slotnick 1983), suggesting olfactory involvement in emotion, complex learning and basic biological functions, respectively. Psychologists were also able to demonstrate remarkably complex learning in rodents: animals trained over time on a series of two-odor discrimination tasks improved their performance accuracy from problem set to problem set (Jennings and Keefer 1969). Their memory for a series of odors is also remarkable: in one study, rats were trained on discrimination tasks with 72 odors and demonstrated almost perfect retention when tested on a random eight-odor subset two weeks later (Slotnick et al. 1991). Conversely, in similar studies of visual discrimination, success in learning and memory remains rather poor, even after extensive training (Hodos 1970). These findings lead to an interesting question: do rodents have a separate system dedicated to formation and recall of olfactory information?

If such a system exists, it would not be remiss to suggest that its activities start with the OB. Evidence exists in the literature for a possible role of the OB in olfactory memory. Infusions of glutamate into the OB have been shown to improve formation, but not recall, of olfactory memories (Rumsey et al. 2001). In contrast, infusions of GABA, the primary neurotransmitter of the olfactory interneurons, reduce the ability of animals to recognize a previously presented odor (Okutani et al. 1999). Olfactory learning is typically a pheromonal phenomenon, but OB involvement has been demonstrated in learning, such as in a ewe's ability to learn the odor of a newborn lamb shortly after birth (Kendrick 1994). Here, there is a significant increase in the number of mitral cells that respond to the odor of the ewe's own lamb. Disruption of noradrenaline from the centrifugal projection from the locus coeruleus in the bulb impairs olfactory learning during the "sensitive period" of olfactory imprinting (the first few hours post-partum; Levy et al. 1990). In terms of the most important neurotransmitters of the bulb, glutamate and GABA, in mice there is a bulbar increase of GABA relative to glutamate during presentation of a learned odor (Brennan et al. 1998). In addition, granule cells show habituation in electrophysiological experiments in which an odor is presented repeatedly (Scott 1991), suggesting that GABAergic neurons may inhibit delivery of information by the mitral cells to the olfactory cortex.

Another body of work on the roles of olfactory neurons involves the use of in vivo imaging, metabolic tracers and electrophysiology to determine their levels of activity during performance of various behaviors. Buonviso and colleagues (1998) gave rats brief exposures to various odors and recorded the number of neurons that responded to the same odors when presented again one or six days later. The number of neurons responding to the now-familiar odors decreased dramatically in both cases. Similar results were seen with expression of the immediate-early gene c-fos, (Montag-Sallaz and Buonviso 2002) which is normally upregulated in the OB within 30 minutes of odor exposure (Baba et al. 1997). An important point of note in these studies is that most of the activity was seen in the granule cell layer. A recent paper by Guthrie and Gall (2003) examined c-fos expression in the rat OB between P0 and P21. They observed that the centers of activity were initially in mitral and tufted cells; however, as the animals aged and neurogenesis progressed, activity shifted to the granule cells, providing further evidence of the importance of these cells in the olfactory circuitry.

In summation, based on the evidence presented here, it seems possible that performance on tasks of olfactory memory may be inversely proportional to the number of interneurons present in the OB, whereas performance on tasks involving olfactory discrimination may improve with the number of interneurons.

Differential roles for new and existing neurons

If, as the evidence would indicate, neural stem cells mediate a series of olfactory functions resident in the bulb, it seems reasonable to suggest that a feedback mechanism from the bulb may regulate neurogenesis at the level of the SVZ. A common hypothesis along these lines is that an increase in amount of olfactory stimuli received would result in a subsequent increase in stem cell-derived neurogenesis. There are many reported instances of extrinsic factors affecting neurogenesis (for example: Shingo et al. 2001; Jin et al. 2002; Parent et al. 2002a,b). However, from a behavioral standpoint, fewer such examples exist in the mammalian literature. Prairie voles are induced into oestrus by olfactory cues upon exposure to an unfamiliar male and exhibit a corresponding increase in proliferation of neuronal precursors in the RMS and anterior SVZ (Smith et al. 2001). In contrast, most of the rodent studies involving olfactory-relevant factors have been less than definitive. Corotto and others (1994) examined the effect of odor deprivation on neurogenesis in mice. They eliminated olfactory input by sealing a naris, which is the external opening of the nasal cavity. Eight weeks later, they examined the extent of proliferation in the RMS and SVZ using BrdU immunohistochemistry. They found that the RMS within and caudal to the OB ipsilateral to the closed naris had significantly fewer BrdU+ cells than in the contralateral, control side. They also noted an increase in number of dying cells in the same regions. However, there were no differences in numbers of proliferating cells in the SVZ proper. More recently, Rochefort and colleagues (2002) performed the reverse experiment: they housed mice for 40 days in an "odor-enriched" environment and gave them injections of BrdU halfway through. Histological analysis revealed that these animals had more BrdU+ cells in their OBs than mice in normal environments, whereas there was no significant difference in number of proliferating cells in the SVZs of either group. Presumably, therefore, an odor-rich environment increases survival of new neurons in the OB but fails to actually increase progenitor numbers. The group went on to suggest that this increased survival of new neurons results in improved memory for odors; however, as the study was purely correlative, these findings are inconclusive.

The findings of increased or decreased cell death in response to modulation of olfactory input, in light of other recent studies, may actually serve to reinforce the notion that the OB demonstrates a level of self-regulation in response to olfactory stimuli. The findings also suggest differential roles for old and new interneurons in the OB. Most olfactory neurogenesis takes place postnatally (Luskin 1993); functional input from olfactory receptor neurons is necessary for their survival, as neonatal granule cells rapidly degenerate on naris closure (Frazier and Brunjes 1988; Frazier-Cierpial and Brunjes 1989; Brunjes 1994). This increase in cell death is not as pronounced when animals are naris-occluded as adults (Fiske and Brunjes 2001). Thus, the new neurons are particularly susceptible to early death. However, even without naris closure, long-term survival analyses of adult-born interneurons show that about 50% of new neurons in the OB die within two months of birth, while the rest survive for as long as 19 months (Kato et al. 2000; Winner et al. 2002; Petreneau and Alvarez-Buylla 2002). Does the OB recruit new neurons for an acute function and eliminate them when they are no longer needed?

Studies conducted in our lab relate this possibility to the olfactory-relevant factors discussed earlier. Recently, we found that pregnancy induces robust production of progenitors from the SVZ, with the number of BrdU-labeled cells increasing by 65% within seven days of gestation (GD7; Shingo et al. 2003). To further determine the fate of these progenitors, we injected mice with BrdU at GD7 and counted the number of BrdU+/NeuN+ neurons in the OB four weeks later, a time by which most of the progenitors have migrated. There were 50% more double-labeled granule cells and 100% more periglomerular cells in mice labeled at GD7 than in virgin controls, indicating the significant recruitment of new neurons to the bulb. We also went on to show that prolactin, a maternal hormone whose expression peaks during the first half of pregnancy (Freeman et al. 2000), is responsible for this surge in neurogenesis. While we performed no behavioral tests in these studies, the results suggest that the recruitment of neurons to the bulb has a particular purpose with respect to olfactory behavior, possibly in enhanced discrimination of the young. We also noted no increases in cell death in the pregnant or control bulbs across the time points examined in this study; given the bulb's ability to regulate its cell numbers, however, we would suspect that a course of apoptosis would eventually return the bulb to its 'normal' state after weaning, pending a repeat of the cycle and future neurogenesis requirements.

Conclusions

It is only one of nature's sardonic twists of fate that marries the inexplicable presence of neural stem cells to the inscrutable phenomenon of olfaction. Neural stem cells pose an unusual biological mystery in their identity, capabilities, functions and purpose. Their functions during development, though not entirely clear, can be imagined without great difficulty. In the adult, however, the questions are more involved: why do granule cells in the OB and hippocampus need replacement? How and when are neural stem cells specified to produce these particular neuronal types? Why is there a paucity of neurons produced elsewhere in the CNS when needed, such as after injury? We have attempted to answer some of these questions in this short chapter; but those answers themselves only raise the many further queries that will occupy neuroscientists for many years to come.

While some evidence points towards the possibility of neurogenesis in regions of the brain other than the OBs, in the adult neocortex to be exact (Magavi et al. 2000), it is hardly a robust phenomenon by default. Some recent attempts to determine the role of the neural stem cell have focused on the stem cell's particular niche as determinant of function (Lim et al. 2000; Hitoshi et al. 2002). This theory assumes that the stem cell is largely or entirely non cell-autonomous in its actions. This may explain the surprising and controversial reports of transdifferentiation from neural to various other fates, such as blood and muscle (Bjornson et al. 1999; Galli et al. 2000, but see Morshead et al. 2002) and accounts for the wide array of receptors and cytokines expressed by a presumptively primitive cell (Quesenberry et al. 1999). Whatever the reason, neural stem cells provide a remarkably useful tool for research into the neurobiology of the brain and have powerful potential in medicine.

References

Acebes A, Ferrús A (2001) Increasing the number of synapses modifies olfactory perception in Drosophila. J Neurosci 21:6264–6273

Baba K, Ikeda M, Houtani T, Nakagawa H, Ueyama T, Sato K, Sakuma S, Yamashita T, Tsukahara Y, Sugimoto T (1997) Odor exposure reveals non-uniform expression profiles of c-Jun protein in rat olfactory bulb neurones. Brain Res 774:142–148

Bjornson CR, Rietze RL, Reynolds BA, Magli MC, Vescovi AL (1999) Turning brain into blood: a hematopoietic fate adopted by adult neural stem cells in vivo. Science 283:534–537

Brennan PA, Schellink HM, De La Riva C, Kendrick KM, Keverne EB (1998) Changes in neurotransmitter release in the olfactory bulb following an olfactory conditioning task in mice. Neuroscience 87:583–590

Brunjes PC (1994) Unilateral naris closure and olfactory system development. Brain Res Brain Res Rev 19:146–160

Buck L, Axel RA (1991) A novel multigene family may encode odorant receptors: a molecular basis for odor recognition. Cell 65:175–187

Buonoviso N, Chaput MA (1990) Response similarity to odors in olfactory bulb neurons presumed to be connected to the same glomerulus: electrophysiological study using single-unit recordings. J Neurophysiol 63:447–454

Buonviso N, Gervais R, Chalansonnet M, Chaput M (1998) Short-lasting exposure to one odor decreases general reactivity in the olfactory bulb of adult rats. Eur J Neurosci 10:2472–2475

Cajal RY (1911) Histologie du Système Nerveux de l'Homme et des Vertébrés. Maloine, Paris

Cattaneo E, McKay R (1990) Proliferation and differentiation of neuronal stem cells regulated by nerve growth factor. Nature 347:762–765

Corotto FS, Henegar JR, Maruniak JA (1994) Odor deprivation leads to reduced neurogenesis and reduced neuronal survival in the olfactory bulb of the adult mouse. Neuroscience 61:739–744

Craig CG, Tropepe V, Morshead CM, Reynolds BA, Weiss S, van der Kooy D (1996) In vivo growth factor expansion of endogenous subependymal neural precursor cell populations in the adult mouse brain. J Neurosci 16:2649–2658

Davis AA, Temple S (1994) A self-renewing multipotential stem cell in embryonic rat cerebral cortex. Nature 372:263–266

Doetsch F, Alvarez-Buylla A (1996) Network of tangential pathways for neuronal migration in the adult mammalian brain. Proc Natl Acad Sci USA 93:14895–14900

Doetsch F, Garcia-Verdugo JM, Alvarez-Buylla A (1999) Regeneration of a germinal layer in the adult mammalian brain. Proc Natl Acad Sci USA 96:11619–11624

Doty RL (2001) Olfaction. Annu Rev Psychol 52:423–452

Dzierzak E, Medvinsky A, de Bruijin M (1998) Qualitative and quantitative aspects of haematopoietic cell development in the mammalian embryo. Immunol Today 19:228–236

Firestein S (2001) How the olfactory system makes sense of scents. Nature 413:211–218

Fiske BK, Brunjes PC (2001) Cell death in the developing and sensory-deprived rat olfactory bulb. J Comp Neurol 431:311–319

Frazier LL, Brunjes PC (1988) Unilateral odor deprivation: early postnatal changes in olfactory bulb cell density and number. J Comp Neurol 269:355–370

Frazier-Cierpial LL, Brunjes PC (1989) Early postnatal differentiation of granule cell dendrites in the olfactory bulbs of normal and unilaterally odor-deprived rats. Brain Res Dev Brain Res 47:129–136

Freeman ME, Kanyicska B, Lerant A, Nagy G (2000) Prolactin: structure, function, and regulation of secretion. Physiol Rev 80:1523–1631

Galli R, Borello U, Gritti A, Minasi MG, Bjornson C, Coletta M, Mora M, De Angelis MG, Fiocco R, Cossu G, Vescovi AL (2000) Skeletal myogenic potential of human and mouse neural stem cells. Nature Neurosci 10:986–991

Garcia-Verdugo JM, Doetsch F, Wichterle H, Lim DA, Alvarez-Buylla A (1998) Architecture and cell types of the adult subventricular zone: in search of the stem cells. J Neurobiol 36:234–248

Gheusi G, Cremer H, McLean H, Chazal G, Vincent JD, Lledo PM (2000) Importance of newly generated neurons in the adult olfactory bulb for odor discrimination. Proc Natl Acad Sci USA 97:1823–1828

Gogos JA, Osborne J, Nemes A, Mendelsohn M, Axel R (2000) Genetic ablation and restoration of the olfactory topographic map. Cell 13:609–620

Guthrie KM, Gall C (2003) Anatomic mapping of neuronal odor responses in the developing rat olfactory bulb. J Comp Neurol 455:56–71

Hadjantonakis A, Papaionnau V (2001) The stem cell of early embryos. Differentiation 68:159–166

Hall PA, Watt FM (1989) Stem cells: the generation and maintenance of cellular diversity. Development 106:619–633

Hammang JP, Archer DR, Duncan ID (1997) Myelination following transplantation of EGF-responsive neural stem cells into a myelin-deficient environment. Exp Neurol 147:84–95

He W, Ingraham C, Rising L, Goderie S, Temple S (2001) Multipotent stem cells from the mouse basal forebrain contribute GABAergic neurons and oligodendrocytes to the cerebral cortex during embryogenesis. J Neurosci 21:8854–8862

Hildebrand JG, Shepherd GM (1997) Mechanisms of olfactory discrimination: Converging evidence for common principles across phyla. Annu Rev Neurosci 20:595–631

Hitoshi S, Tropepe V, Ekker M, van der Kooy D (2002) Neural stem cell lineages are regionally specified, but not committed, within distinct compartments of the developing brain. Development 129:233–244

Hodos W (1970) Evolutionary interpretation of neural and behavioral studies of living vertebrates. In: Schmitt FO (ed) The Neurosciences: 2nd Study Program. New York, pp 26–39

Hosler JS, Buxton KL, Smith BH (2000) Impairment of olfactory discrimination by blockade of GABA and nitric oxide activity in the honeybee antennal lobes. Behav Neurosci 114:514–525

Hu H, Tomasiewicz H, Magnuson T, Rutishauser U (1996) The role of polysialic acid in the migration of olfactory bulb interneuron precursors in the subventricular zone. Neuron 16:735–743

Huttner WB, Brand M (1997) Asymmetric division and polarity of neuroepithelial cells. Curr Opin Neurobiol 7:29–39

Jackowski A, Parnavelas JG, Lieberman AR (1978) The reciprocal synapse in the external plexiform layer of the mammalian olfactory bulb. Brain Res 159:17–28

Jahr CE, Nicoll RA (1982) An intracellular analysis of dendrodendritic inhibition in the turtle in vitro olfactory bulb. J Physiol 326:213–234

Jennings JW, Keefer LH (1969) Olfactory learning set in two varieties of domestic rat. Physiol Rep 24:3–15

Jin K, Zhu Y, Sun Y, Mao XO, Xie L, Greenberg DA (2002) Vascular endothelial growth factor (VEGF) stimulates neurogenesis in vitro and in vivo. Proc Natl Acad Sci USA 99:11946–11950

Kalyani AJ, Piper D, Mujtaba T, Lucero MT, Rao MS (1998) Spinal cord neuronal precursors generate multiple neuronal phenotypes in culture. J Neurosci 18:7856–7868

Kaplan MS, McNelly NA, Hinds JW (1985) Population dynamics of adult formed granule neurons of the rat olfactory bulb. J Comp Neurol 239:117–125

Kato T, Yokcrouchi K, Kawagishi K, Fushima N, Miwa T, Moriizumi T (2000) Fate of newly formed periglomerular cells in the olfactory bulb. Acta Otolaryngol 120:876–879

Kauer JS (1974) Response patterns of amphibian olfactory bulb neurons to odor stimulation. Brain Res 188:139–154

Kendrick KM (1994) Neurobiological correlates of visual and olfactory recognition in sheep. Behav Processes 33:89–112

Kilpatrick TJ, Bartlett PF (1995) Cloned multipotential precursors from the mouse cerebrum require FGF-2, whereas glial restricted precursors are stimulated with either FGF-2 or EGF. J Neurosci 15:3653–3661

Lajtha LG (1979) Stem cell concepts. Differentiation 14:23–34

Leon M (1992) The neurobiology of filial learning. Annu Rev Psychol 43:377–398

Leveteau J, MacLeod P (1966) Olfactory discrimination in the rabbit olfactory glomerulus. Science 155:175–176

Levy F,Gervais R, Kindermann U, Orgeur P, Piketty V (1990) Importance of beta-adrenergic receptors in the olfactory bulb of sheep for recognition of lambs. Behav Neurosci 104:464–469

Lillien L, Raphael H (2000) BMP and FGF regulate the development of EGF-responsive neural progenitor cells. Development 127:4993–5005

Lim DA, Tramontin AD, Trevejo JM, Herrera DG, Garcia-Verdugo JM, Alvarez-Buylla A (2000) Noggin antagonizes BMP signalling to create a niche for adult neurogenesis. Neuron 28: 713–726

Lois C, Garcia-Verdugo JM, Alvarez-Buylla A (1996) Chain migration of neuronal precursors. Science 271:978–981

Luskin MB (1993) Restricted proliferation and migration of postnatally generated neurons derived from the forebrain subventricular zone. Neuron 11:173–189

Luskin MB, Zigova T, Soteres BJ, Stewart RR (1997) Neuronal progenitor cells derived from the anterior subventricular zone of the neonatal rat forebrain continue to proliferate in vitro and express a neuronal phenotype. Mol Cell Neurosci 8:351–366

Magavi SS, Leavitt BR, Macklis JD (2000) Induction of neurogenesis in the neocortex of adult mice. Nature 405:951–955

Marshak DR, Gottlieb D, Gardner RL (2001) Introduction: Stem cell biology. In: Marshak DR, Gardner RL, Gottlieb D (eds) Stem cell biology. Cold Spring Harbor, New York, pp 1–16

McConnell SK (1995) Constructing the cerebral cortex: neurogenesis and fate determination. Neuron 15:761–768

Meredith M (1986) Patterned response to odor in mammalian olfactory bulb: the influence of intensity. J Neurophysiol 56:572–597

Mombaerts P, Wang F, Dulac C, Chao SK, Nemes A, Mendelsohn M, Edmondson J, Axel R (1996) Visualizing an olfactory sensory map. Cell 87:675–686

Montag-Sallaz M, Buonviso N (2002) Altered odor-induced expression of c-fos and arg 3.1 immediate early genes in the olfactory system after familiarization with an odor. J Neurobiol 52:61–72

Monti-Graziadei GA, Graziadei PPC (1979) Neurogenesis and neuron regeneration in the olfactory system of mammals II: Degeneration and reconstitution of the olfactory sensory neurons after axotomy. J Neurocytol 8:187–213

Mori K (1987) Membrane and synaptic properties of identified neurons in the olfactory bulb. Prog Neurobiol 29:275–320

Morrison SJ, Shah NM, Anderson DJ (1997) Regulatory mechanisms in stem cell biology. Cell 88: 287–298

Morshead CM, Reynolds BA, Craig CG, McBurney MW, Staines W A, Morassutti D, Weiss S, van der Kooy D (1994) Neural stem cells in the adult mammalian forebrain: a relatively quiescent subpopulation of subependymal cells. Neuron 13:1071–1082

Morshead CM, Benveniste P, Iscove NN, van der Kooy D (2002) Hematopoietic competence is a rare property of neural stem cells that may depend on genetic and epigenetic alterations. Nature Med 3:268–273

Okutani F, Yagi F, Kaba H (1999) GABAergic control of olfactory learning in young rats. Neuroscience 93:1297–1300

Palmer TD, Takahashi J, Gage FH (1997) The adult rat hippocampus contains primordial neural stem cells. Mol Cell Neurosci 8:389–404

Parent JM, Valentin VV, Lowenstein DH (2002a) Prolonged seizures increase proliferating neuroblasts in the adult rat subventricular zone-olfactory bulb pathway. J Neurosci 22: 3174–3188

Parent JM, Vexler ZS, Gong C, Derugin N, Ferriero DM (2002b) Rat forebrain neurogenesis and striatal neuron replacement after focal stroke. Ann Neurol 52:802–813

Pencea V, Bingaman KD, Freedman LJ, Luskin MB (2001) Neurogenesis in the subventricular zone and the rostral migratory stream of the neonatal and adult primate forebrain. Exp Neurol 172: 1–16

Petreneau L, Alvarez-Buylla A (2002) Maturation and death of adult-born olfactory bulb granule neurons: Role of olfaction. J Neurosci 22:6106–6113

Porter RH, Etscorn F (1975) A primacy effect for olfactory imprinting in spiny mice (Acomys cahirinus). Behav Biol 15:511–517

Potten CS, Loeffler M (1990) Stem cells: attributes, cycles, spirals, pitfalls and uncertainties. Lessons for and from the crypt. Development 110:1001–1020

Powell TP, Cowan WM, Raisman G (1965) The central olfactory connexions. J Anat 99:791–793

Price J (2001) Neural stem cells – where are you? Nat Med 7:998–999

Price JL, Powell TPS (1970) The morphology of granule cells in the olfactory bulb. J Cell Sci 7: 91–123

Price JL, Slotnick BM (1983) Dual olfactory representation in the rat thalamus: an anatomical and electrophysiological study. J Comp Neurol 215:63–77

Quesenberry PJ, Hulspas R, Joly M, Benoit B, Engstrom C, Rielly J, Savarese T, Pang L, Recht L, Ross A, Stein G, Stewart M (1999) Correlates between hematopoiesis and neuropoiesis: neural stem cells. J Neurotrauma 16:661–666

Rakic P (1995) A small step for the cell, a giant leap for mankind: a hypothesis of neocortical expansion during evolution. Trends Neurosci 18:383–388

Represa A, Shimazaki T, Simmonds M, Weiss S (2001) EGF-responsive stem cells are a transient population in the developing mouse spinal cord. Eur J Neurosci 14:452–462

Ressler KJ, Sullivan SL, Buck LB (1994) Information coding in the olfactory system: evidence for a stereotyped and highly organized epitope map in the olfactory bulb. Cell 79:1245–1255

Reyher CK, Lubke J, Larsen WJ, Hendrix GM, Shipley MT, Baumgarten HG (1991) Olfactory bulb granule cell aggregates: morphological evidence for intraperikaryal electronic coupling via gap junctions. J Neurosci 11:1485–1495

Reynolds BA, Weiss S (1992) Generation of neurons and astrocytes from isolated cells of the adult mammalian central nervous system. Science 255:1707–1710

Reynolds BA, Weiss S (1996) Clonal and population analyses reveal that an EGF-responsive mammalian embryonic CNS precursor is a stem cell. Dev Biol 175:1–13

Reynolds BA, Tetzlaff W, Weiss S (1992) A multipotent EGF-responsive striatal embryonic progenitor cell produces neurons and astrocytes. J Neurosci 12:4565–4574

Rochefort C, Gheusi G, Vincent JD, Lledo PM (2002) Enriched odor exposure increases the number of newborn neurons in the adult olfactory bulb and improves odor memory. J Neurosci 22: 2679–2689

Rumsey JD, Darby-King A, Harley CW, McLean JH (2001) Infusion of the metabotropic receptor agonist, DCG-IV, into the main olfactory bulb induces olfactory preference learning in rat pups. Brain Res Dev Brain Res 128:177–179

Schneider SP, Macrides F (1978) Laminar distribution of interneurons in the mail olfactory bulb of the adult hamster. Brain Res Bull 3:73–82

Schoppa NE, Kinzie JM, Sahara Y, Segerson TP, Westbrook GL (1998) Dendrodendritic inhibition in the olfactory bulb is driven by NMDA receptors. J Neurosci 18:6790–6802

Scott JW (1991) Central processing of olfaction. J Steroid Biochem Mol Biol 39:593–600

Shepherd GM, Greer CA (1990) Olfactory bulb. In: Shepherd GM (ed) The synaptic organization of the brain. Oxford University Press, New York, pp 133–169

Shingo T, Sorokan ST, Shimazaki T, Weiss S (2001) Erythropoietin regulates the in vitro and in vivo production of neuronal progenitors by mammalian forebrain neural stem cells. J Neurosci 21: 9733–9743

Shingo T, Gregg C, Enwere E, Fujikawa H, Hassam R, Geary C, Cross JC, Weiss S (2003) Pregnancy-stimulated neurogenesis in the adult female forebrain mediated by prolactin. Science 299: 117–120

Shipley MT, Ennis M (1996) Functional organization of olfactory system. J Neurobiol 30:123–176

Slotnick BM, Kujera A, Silberberg AM (1991) Olfactory learning and odor memory in the rat. Physiol Behav 50:555–561

Smith MT, Pencea V, Wang Z, Luskin MB, Insel TR (2001) Increased number of BrdU-labeled neurons in the rostral migratory stream of the estrous prairie vole. Horm Behav 39:11–21

Steinbrecht RA (1969) Comparative morphology of olfactory receptors. In: Pfaffmann C (ed) Olfaction and taste III. Rockfeller University Press, New York, pp 3–21

Temple S (2001) The development of neural stem cells. Nature 414:112–127

Tomasiewicz H, Ono K, Yee D, Thompson C, Goridis C, Rutishauser U, Magnuson T (1993) Genetic deletion of a neural cell-adhesion molecule variant (NCAM-180) produces distinct defects in the central nervous system. Neuron 11:1163–1174

Treloar HB, Feinstein P, Mombaerts P, Greer CA (2002) Specificity of glomerular targeting by olfactory sensory axons. J Neurosci 22:2469–2477

Tropepe V, Sibilia M, Ciruna BG, Rossant J, Wagner EF, van der Kooy D (1999) Distinct neural stem cells proliferate in response to EGF and FGF in the developing mouse telencephalon. Dev Biol 208:166–188

van der Kooy D, Weiss S (2000) Why stem cells? Science 287:1439–1442

Vescovi AL, Reynolds BA, Fraser DD, Weiss S (1993) bFGF regulates the proliferative fate of unipotent (neuronal) and bipotent (neuronal/astroglial) EGF-generated CNS progenitor cells. Neuron 11:951–966

Wanaka A, Milbrandt J, Johnson EM Jr (1991) Expression of FGF receptor gene in rat development. Development 111:455–468

Wellis DP, Scott JW, Harrison TA. (1989) Discrimination among odorants by single neurons of the rat olfactory bulb. J Neurophysiol. 61:1161–1177

Weissman IL (2000) Stem cells: units of development, units of regeneration, and units of evolution. Cell 100:157–168

Wilson DA, Leon M (1987) Evidence of lateral synaptic interactions in olfactory bulb output cell responses to odors. Brain Res 417:175–180

Winkler C, Fricker RA, Gates MA, Olsson M, Hammang JP, Carpenter MK, Björklund A (1998) Incorporation and glial differentiation of mouse EGF-responsive neural progenitor cells after transplantation into the embryonic rat brain. Mol Cell Neurosci 11:99–116

Winner B, Cooper-Kuhn CM, Aigner R, Jürgen W, Kuhn HG (2002) Long-term survival and cell death of newly generated neurons in the adult rat olfactory bulb. Eur J Neurosci 16:1681–1689

Converting ES Cell into Neurons

A. Smith[1]

Summary

Our laboratory is studying determination of neuroectoderm from pluripotent embryonic stem (ES) cells. We have shown that, in defined medium, ES cells can enter directly into the neural lineage, bypassing the requirement for multicellular aggregation or treatment with retinoic acid. The process by which pluripotent ES cells acquire neural identity in adherent culture can be visualised and recorded at the level of individual colonies. Commitment to neuroectoderm is promoted by fibroblast growth factor-4 and is suppressed by serum, Wnts, bone morphogenetic proteins and extracellular matrix. The cellular and molecular dissection of ES cell lineage choice may yield insights into the mechanism underlying neural determination in the mammalian embryo. This simple culture system is also a step towards the "directed," homogeneous differentiation of ES cells required for biopharmaceutical and clinical applications.

Introduction

Embryonic stem (ES) cells are permanent cell lines established from pre-implantation mouse embryos without the use of any transforming or immortalising agent (Evans and Kaufman 1981; Martin 1981). These cells can be expanded without limit in vitro whilst retaining the capacity for differentiation into all lineages and mature cell types of the animal. Furthermore, if ES cells are reintroduced into the embryo, they colonise all tissues of the developing foetus, giving rise to chimaeric animals with fully functional terminal cell types derived from the cultured stem cells (Bradley et al. 1984; Nagy et al. 1993). ES cells therefore constitute a unique tool for basic research in developmental genetics and, in particular, for discovering how distinct cell lineages diversify from a common, pluripotent founder cell (Smith 2001). Furthermore, the advent of counterpart pluripotent cells derived from human embryos (Thomson et al. 1998) creates the potential for major biomedical applications (Fig. 1). The

[1] Institute for Stem Cell Research, University of Edinburgh, King's Buildings, West Mains Road, Edinburgh, EH9 3JQ, Scotland, Mail: Austin.smith@ed.ac.uk

Gage et al.
Stem Cells in the Nervous System:
Functional and Clinical Implications
© Springer-Verlag Berlin Heidelberg 2004

Stem Cell Medicine

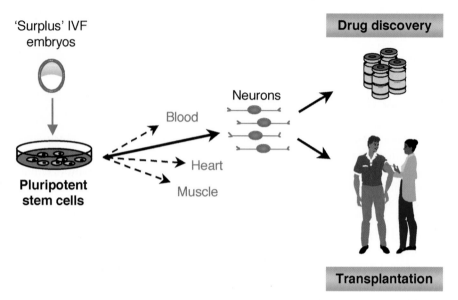

'Surplus' IVF embryos

Drug discovery

Neurons

Blood

Pluripotent stem cells

Heart

Muscle

Transplantation

Fig. 1. Stem cell medicine.Pluripotent stem cells derived from donated supernumerary embryos could provide a limitless supply of differentiated cell types for use in pharmaceutical discovery and cell transplantation therapy. A key requirement for such applications will be the ability to channel stem cell differentiation into efficient production of specific cell types.

opportunity for generating large numbers of specific, differentiated progeny could be exploited, both to provide new assays for drug development (Gorba and Allsopp 2003) and to produce cells and tissue for transplantation therapy (Smith 1998; Solter and Gearhart 1999).

It is well established that ES cells can produce a diversity of morphologically and molecularly differentiated cell types in culture (Keller 1995; Smith 2001). However, questions remain concerning the fidelity of the differentiation. For tissue repair, the critical issue is whether these in vitro-generated cells can integrate and function when transplanted into adult tissue. Grafts of ES cell-derived cardiomyocytes into ischaemic rat hearts provided the first indication that this could be the case (Klug et al. 1996). Subsequently encouraging results came from transplants of glial precursors derived from mouse ES cells into a rat model of acute demyelination (Brustle et al. 1999). The most persuasive evidence to date is provided by two recent transplantation studies in Parkinsonian rats (Bjorklund et al. 2002; Kim et al. 2002). In both cases, appreciable numbers of graft-derived tyrosine hydroxylase-positive cells were demonstrated, and significant functional recovery was apparent in the amphetamine-induced rotation assay and other motor tests.

Thus it appears that correctly specified adult cell types can be generated from ES cells without passing through the embryo. But must ES cells obey an invariant sequence of developmental fate choices derived from embryonic development or can they arrive at terminal specification via alternative pathways, unconstrained by embryology? In their studies, Kim et al. (2002) endeavoured to recapitulate the embryonic specification of dopaminergic neurons in vitro via a multi-stage differentiation process, including genetic manipulation. They then transplanted cell populations enriched for immature neurons of the mesencephalic dopaminergic phenotype. In contrast, Bjorklund et al. (2002) injected naïve ES cells directly into the brain, relying on the host environment to stimulate dopaminergic differentiation directly. They found that the adult brain promotes dopaminergic differentiation of ES cells quite efficiently. It is unclear how this is achieved, particularly since expanded neural stem cells generally show very limited dopaminergic capacity in vivo (Dunnett and Bjorklund 1999). Nonetheless the finding does suggest that it may not be necessary or even advantageous to direct ES cell differentiation in vitro to the point of terminal specification. However, transplantation of undifferentiated ES cells is not a clinical option, because in this state the cells are inherently proliferative and tumourigenic (Burdon et al. 2002) and consequently generate teratomas and teratocarcinomas (Brustle et al. 1997; Deacon et al. 1998).

Thus, how far along a particular developmental avenue ES cells should be pushed prior to transplantation remains an open question. On the other hand, for biopharmaceutical applications, such as screening for neuroprotective agents, terminally differentiated phenotypes are clearly of most interest. Consequently, there is a need to develop optimal procedures for and understanding of primary ES cell commitment and lineage-specific differentiation. We cannot yet claim an ability to "direct" ES cell differentiation quantitatively. Some progress has been made with controlling intermediate stages of lineage progression, as exemplified by Kim et al. (2002), but mastering the full sequence of steps in the generation of any particular terminally differentiated phenotype remains elusive. There are two major challenges: first, to understand and manipulate lineage choices; second, to develop culture conditions that support the viability and maturation of progenitor and terminal phenotypes in vitro. Our laboratory has been focussing on the first issue, with particular reference to conversion of ES cells into neural-restricted precursors.

Sox1-gfp Reporter of Neural Determination

To investigate in more detail the pathway by which ES cells acquire neural specification, we developed a reporter cell line in which expression of the *Sox1* gene can be monitored. Sox1 is a member of the B group of Sox family transcription factors (Pevny and Lovell-Badge 1997). We chose Sox1 as a marker due to the following expression features:

- activation specifically in neuroectoderm at formation of the neural plate (Wood and Episkopou 1999).
- expression throughout entire neuraxis, encompassing all regionally-restricted neural progenitors (Pevny et al. 1998)
- down-regulation at the onset of both neuronal and glial differentiation (Pevny et al. 1998)
- no foetal expression outwith the CNS except for the lens (Pevny et al. 1998)
- no expression in undifferentiated ES cells, but present in ES cell-derived neural precursors (Li et al. 1998)

Homologous recombination was used to introduce eGFP into the Sox1 gene (Aubert et al. 2003). The resultant 46C ES cells do not express detectable GFP until induced to differentiate under conditions that promote neural fates (Aubert et al. 2002). The GFP reporter can be used to visualise living neural precursor cells by fluorescence microscopy, to quantitate them by flow cytometry analysis, and to purify them by fluorescence-activated cell sorting (FACS; Stavridis and Smith 2003). This can be applied to identify and isolate neuroepithelial cells generated by in vitro differentiation of ES cells and also in the foetal and adult animal, using transgenic mice produced from the 46C ES cells (Aubert et al. 2003).

Neural Determination of ES Cells in Serum-Free Monolayer Culture

Previous protocols for generation of neurons and glia from ES cells have relied on aggregation and either treatment with retinoic acid (Bain et al. 1995) or replating in selective minimal media in which the majority of non-neural cells cannot survive (Okabe et al. 1996). Recently a protocol was described for efficient induction of neurons from ES cells via co-culture with a particular stromal cell line, PA6 (Kawasaki et al. 2000). The effect of PA6 cells is ascribed to an uncharacterised stromal cell-derived inducing activity (SDIA). It has also been reported that ES cells placed in serum-free suspension culture will spontaneously develop into aggregates of neural stem cells, "neurospheres," at low frequency (Tropepe et al. 2001). It is not clear whether this represents de novo differentation or selection of rare, pre-differentiated neural precursors within the ES cell culture. The former interpretation would be consistent with the "default" model of neural fate determination, which states that neural fates are acquired in the absence of any instructive signals (Munoz-Sanjuan and Brivanlou 2002).

To examine the "default" proposition further, we plated 46C ES cells in serum-free medium and withdrew the self-renewal cytokine leukaemia inhibitory factor (LIF; Burdon et al. 2002) to trigger differentiation. In contrast to the situation in serum-containing medium, where ES cells differentiate into non-neural cell types and *Sox1-gfp* is not activated, in the absence of serum, GFP becomes detectable in the majority of cells within 96 hours (Ying et al. 2003). Subsequently these cultures produce large numbers of neurons. Glial cells, both astrocytes and oligodendrocytes, also emerge, particularly if the GFP-positive cells are exposed to basic fibroblast growth factor (FGF-2) or serum.

The conversion of ES cells into Sox1-GFP expressing, nestin-positive, neural precursor cells appears to be quantitative and is not dependent on selective cell death. Throughout the period in which Sox1-GFP cells are being generated, cell numbers increase exponentially, with no appreciable lag phase after plating. Direct observation of individual colonies confirms that cells proliferate continuously and convert from GFP-negative to GFP-positive without significant cell elimination (Ying et al. 2003).

An important consideration in any study using ES cells is the generality of any effect. ES cells are prone to epigenetic and genetic instability, such that deviant phenotypes can readily arise. The 46C ES cells pass the test of chimaera contribution and germline transmission. In addition we have examined a variety of different ES cells and found that all undergo neural differentiation using this serum-free protocol. It should be noted, however, that in our laboratory ES cells are propagated using LIF alone (Smith 1991) and not by co-culture with feeder cells. The presence of feeder cells is likely to disrupt the differentiation process.

The serum-free medium composition contains only transferrin, albumin and insulin as protein components (full details are given in Ying and Smith 2003). Of these, transferrin is essential for cell viability. Albumin can be dispensed with at some compromise of plating efficiency but without reducing the rate or frequency of conversion to GFP positivity. Insulin may be substituted for by insulin-like growth factor I (IGF-I), indicating that its effects are mediated via the IGF receptor. In the absence of both albumin and insulin (or IGF-I), cell survival is compromised, and the proportion of GFP-positive cells is reduced to around 30%. This is likely to be at least in part because insulin/IGF-I is a survival factor for neural precursors, although IGF-1 may also directly promote neural induction, as suggested by studies in Xenopus (Pera et al. 2001).

Fibroblast Growth Factor Signalling Appears to Drive Neural Determination

The observation that neural specification proceeds in media in which the sole protein component is transferrin seems consistent with the notion of a "default" programme. However, this does not take account of the autocrine and paracrine effects of factors expressed within the ES cell cultures. Particularly relevant is the expression by ES cells of a soluble member of the fibroblast growth factor family, FGF-4 (Wilder et al. 1997). Studies in Xenopus (Launay et al. 1996) and chick (Streit et al. 2000; Wilson et al. 2000) have implicated FGF signalling in competence for neural induction in vertebrates (Wilson and Edlund 2001). Whether FGF signalling is involved in neural determination in the mammalian embryo is currently unclear, as *Fgf4* and *Fgfr2* null embryos do not develop beyond implantation (Feldman et al. 1995; Arman et al. 1998).

We examined the effect of adding FGF-4 to the ES cell cultures and found that it does enhance the initial rate of neural conversion. The effect is modest, possibly because the levels of endogenous FGF-4 are near saturating. Significantly, however, FGF-2, which is the standard mitogen used to expand neural progenitor cells, does not increase the appearance of Sox1-GFP-positive cells. This finding suggests that the effect of FGF-4 is specifically on the process of primary neural conversion rather than subsequent expansion of the neural precursor population. We also investigated the consequences of blocking FGF receptor signalling. The pharmacological inhibitor of FGF receptor tyrosine kinase activity SU5402 (Mohammadi et al. 1997) significantly suppresses the emergence of Sox1-GFP-positive cells (Ying et al. 2003). Furthermore, ES cells transfected with a truncated, dominant-negative FGF receptor are completely blocked in neural differentiation (Ying et al. 2003). These data suggest that autocrine FGF-4 primes ES cells for entry into the neuroectodermal lineage on withdrawal of self-renewal stimuli (Fig. 2). Analysis of FGF-4-negative ES cells is required to provide formal proof of this argument, however, and to determine whether other FGFs may also contribute. Furthermore it is possible that other factors expressed within the ES cell cultures, for example, IGFs, may play a stimulatory role.

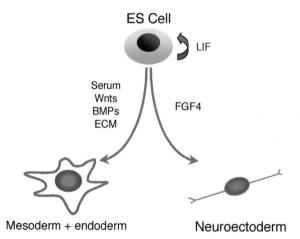

Autocrine FGF4 Primes ES Cells for Neural Differentiation

ES Cell

LIF

Serum
Wnts
BMPs
ECM

FGF4

Mesoderm + endoderm

Neuroectoderm

Fig. 2. ES cell lineage choice. On withdrawal of LIF, ES cells exit self-renewal and enter a differentiation pathway. The decision between alternative differentiated lineages is influenced by extrinsic sig-nals. Factors such as BMPs, Wnts and laminin direct non-neural differentiation. In the absence of these factors, FGF4, which is expressed by ES cells, promotes neural lineage determination.

Anti-Neural Factors in ES Cell Lineage Choice

As noted above, Sox1-GFP expression is not seen in adherent cultures in serum-containing medium and neurons do not develop. FGF-4 is still expressed under these conditions, however. The implication is that serum overrides FGF signalling and directs ES cells into alternative differentiation pathways. In Xenopus, it has been well documented that BMPs act as anti-neural factors (Harland 2000). Previous studies with ES cells had indicated that BMPs suppressed neural differentiation (Tropepe et al. 2001) and promoted mesodermal fates (Johansson and Wiles 1995). Consistent with this finding, we found that the addition of BMP4 completely blocked the emergence of Sox1-GFP-positive cells and directed differentiation of large, flattened cells of indeterminate phenotype (Ying et al. 2003).

A third factor that can influence the direction of ES cell differentiation is the extracellular matrix substrate. ES cells are routinely cultured on plastic coated with denatured collagen (gelatin), which improves their adherence to the surface (Smith 1991). In contrast, neurons are generally cultured on laminin-coated dishes. For neural determination of ES cells, it is important to maintain the cells on gelatin. We found that laminin markedly suppresses neural determination and, like BMP, induces differentiation into large, flat cells. Fibronectin also reduces neural determination, although to a lesser extent than laminin. A possible explanation for this effect is that activation of integrin signalling by matrix components can influence ES cell lineage choice. Alternatively, the matrix components may be contaminated with trace amounts of BMPs or other factors. One consequence of the use of gelatin is that it is not an ideal substrate for neuronal differentiation. Accordingly, for optimal generation of neurons, it is advisable to replate the cultures after several days onto laminin-coated dishes (Ying and Smith 2003).

Finally, in a differential screen for genes that promote neural differentiation of ES cells, we identified that the Wnt antagonist secreted frizzled-like protein-2 (sFRP2) (Aubert et al. 2002). sFRP2 is activated during retinoic acid-induced neural differentiation and is also expressed selectively in neural tissues in the early embryo. Forced expression sFRP2 promotes neural differentiation in aggregates independent of retinoic acid treatment and accelerates neural determination in the adherent serum-free protocol (Aubert et al. 2002 and unpublished data). sFRPs act by extracellular sequestration of Wnt proteins and are not thought to have any intrinsic signalling capacity. The interpretation that sFRP2 acts by restricting anti-neural activity of Wnts was substantiated by forced expression on Wnt1 in ES cells, which suppressed neural differentiation. Thus Wnts also function as anti-neural factors in ES cell lineage choice.

Overall, therefore, a variety of factors appear able to instruct ES cells into non-neural lineages at the expense of neural determination (Fig. 2). Removal or blockade of these extrinsic components is thus a key requirement for steering ES cells towards a neuroectodermal fate.

Current Perspectives and Future Prospects

The overall conversion of ES cells to neural precursors in the serum-free monolayer system ranges between 50-80%. Reproducibly, around 10% of the ES cells remain undifferentiated stem cells, whilst variable proportions of cells differentiate along non-neural pathways. It is reasonable to ask why differentiation is not uni-directional and 100% penetrant, when in principle all of the cells are in an identical environment. Inductive factor(s) expressed by the ES cells may promote non-neural commitment. Obvious candidates for anti-neural agents are BMPs. Perhaps surprisingly, however, the addition of BMP antagonists noggin or chordin does not enhance neural commitment in this system (Ying et al. 2003), indicating that endogenous BMP levels are insufficient to direct differentiation. On the other hand, FGF4 does not seem to be limiting, except at very low cell densities, since addition of FGF4 does not significantly enhance neural determination in moderate- or high-density cultures. Autocrine expression of LIF by ES cells (Rathjen et al. 1990) very likely supports the persistence of a fraction of undifferentiated stem cells that may subsequently expand due to production of LIF or related cytokines by neighbouring neural cells. But why do some ES cells escape differentiation initially? Time lapse monitoring reveals that clusters of undifferentiated ES cells persist surrounded by and in direct contact with cells undergoing neural differentiation. Furthermore, it is evident that Sox1-GFP-positive cells emerge asynchronously throughout the culture. One possible explanation could be the heterogeneity in the responsiveness of ES cells to FGF4, for example, due to variability in receptor or co-receptor expression levels. More generally, there may be a stochastic principle underlying the commitment decision. If this is the case, then uniform differentiation may be achievable most readily via a combination of extrinsic instruction with forced expression of lineage-specific master transcription factor(s). An interesting candidate for this approach in neuroectoderm determination is Sox1 itself, which has been reported to promote neural differentiation of P19 embryonal carcinoma cells (Pevny et al. 1998).

A further issue in ES cell differentiation concerns the primary regional identity of the neural precursors and whether this identity can be modulated by extrinsic signals. Compared with the cellular complexity of embryoid body systems, the monolayer system may offer enhanced opportunities to impose anterior or posterior identity due to the direct access of added instructive stimuli to the neural precursors. A further motivation for developing an adherent serum-free protocol was to avoid use of retinoic acid, which is a potent posteriorizing agent. Indeed a recent elegant study by Jessell and co-workers has shown that neural precursors generated by retinoic acid treatment of ES cell aggregates have a spinal cord specification (Wichterle et al. 2002). In contrast, dopaminergic neurons can readily be obtained from monolayer neural precursors by treatment with sonic hedgehog plus FGF-2 (Ye et al. 1998). This finding suggests that some fraction of the cells should have ventral midbrain character. We are currently examining the expression of anterior neural markers to confirm this point and to establish whether the monolayer precursors have a common regional specification or are heterogeneous.

Biomedical applications of ES cells will require efficient commitment to a primary lineage of choice under defined, serum-free culture conditions (Smith 2001). We have demonstrated some progress towards this goal in the case of neural determination of mouse ES cells. Whether the same strategy will be successful with human pluripotent stem cells remains to be determined.

Acknowledgements

Research in the author's laboratory is supported by the Medical Research Council and the Biotechnology and Biological Sciences Research Council of the United Kingdom, and by the Human Frontiers Science Program Organisation. The author is a Medical Research Council Research Professor.

References

Arman E, Krausz-Haffner R, Chen Y, Heath JK, Lonai P (1998) Targeted disruption of fibroblast growth factor (FGF) receptor 2 suggests a role for FGF signaling in pregastrulation mammalian development. Proc Natl Acad Sci USA 95: 5082–5087

Aubert J, Dunstan H, Chambers I, Smith A (2002) Functional gene screening in embryonic stem cells implicates Wnt antagonism in neural differentiation. Nature Biotechnol 20: 1240–1245

Aubert J, Stavridis M, Tweedie S, O'Reilly M, Vierlinger K, Li M, Ghazal P, Pratt T, Mason JO, Roy D, Smith A (2003) Screening for mammalian neural genes via FACS purification of neural precursors from Sox1-gfp knock-in mice. Proc Natl Acad Sci USA, in press

Bain G, Kitchens D, Yao M, Huettner JE, Gottlieb DI (1995) Embryonic stem cells express neuronal properties in vitro. Dev Biol 168:342–357

Bjorklund LM, Sanchez-Pernaute R, Chung S, Andersson T, Chen IY, McNaught KS, Brownell AL, Jenkins BG, Wahlestedt C, Kim KS, Isacson O (2002) Embryonic stem cells develop into functional dopaminergic neurons after transplantation in a Parkinson rat model. Proc Natl Acad Sci USA 99: 2344–2349

Bradley A, Evans MJ, Kaufman MH, Robertson E (1984) Formation of germ-line chimaeras from embryo-derived teratocarcinoma cell lines. Nature 309: 255–256

Brustle O, Spiro AC, Karram K, Choudhary K, Okabe S, McKay RD (1997) In vitro-generated neural precursors participate in mammalian brain development. Proc Natl Acad Sci USA 94: 14809–14814

Brustle O, Jones KN, Learish RD, Karram K, Choudhary K, Wiestler OD, Duncan ID, McKay RD (1999) Embryonic stem cell-derived glial precursors: a source of myelinating transplants. Science 285: 754–756

Burdon T, Smith A, Savatier P (2002) Signalling, cell cycle and pluripotency in embryonic stem cells. Trends Cell Biol 12: 432–438

Deacon T, Dinsmore J, Costantini LC, Ratliff J, Isacson O (1998) Blastula-stage stem cells can differentiate into dopaminergic and serotonergic neurons after transplantation. Exp Neurol 149: 28–41

Dunnett SB, Bjorklund A (1999) Prospects for new restorative and neuroprotective treatments in Parkinson's disease. Nature 399: A32–39

Evans MJ, Kaufman M (1981) Establishment in culture of pluripotential cells from mouse embryos. Nature 292: 154–156

Feldman B, Poueymirou W, Papaioannou VE, DeChiara TM, Goldfarb M (1995) Requirement of FGF-4 for postimplantation mouse development. Science 267: 246–249

Gorba T, Allsopp T (2003) Pharmacological potential of embryonic stem cells. Pharmacol Res47: 269–278

Harland R (2000) Neural induction. Curr Opin Genet Dev 10: 357–362

Johansson BM, Wiles MV (1995) Evidence for involvement of activin A and bone morphogenetic protein 4 in mammalian mesoderm and hematopoietic development. Mol Cell Biol 15: 141–151

Kawasaki H, Mizuseki K, Nishikawa S, Kaneko S, Kuwana Y, Nakanishi S, Nishikawa SI, Sasai Y (2000) Induction of midbrain dopaminergic neurons from ES cells by stromal cell-derived inducing activity. Neuron 28: 31–40

Keller GM (1995) In vitro differentiation of embryonic stem cells. Curr Opin Cell Biol 7: 862–869

Kim JH, Auerbach JM, Rodriguez-Gomez JA, Velasco I, Gavin D, Lumelsky N, Lee SH, Nguyen J, Sanchez-Pernaute R, Bankiewicz K, McKay R (2002) Dopamine neurons derived from embryonic stem cells function in an animal model of Parkinson's disease. Nature 418: 50–56

Klug MG, Soonpaa MH, Koh GY, Field LJ (1996) Genetically selected cardiomyocytes from differentiating embryonic stem cells form stable intracardiac grafts. J Clin Invest 98: 216–224

Launay C, Fromentoux V, Shi DL, Boucaut JC (1996) A truncated FGF receptor blocks neural induction by endogenous Xenopus inducers. Development 122:869–880

Li M, Pevny L, Lovell-Badge R, Smith A (1998) Generation of purified neural precursors from embryonic stem cells by lineage selection. Curr Biol 8: 971–974

Martin GR (1981) Isolation of a pluripotent cell line from early mouse embryos cultured in medium conditioned by teratocarcinoma stem cells. Proc Natl Acad Sci USA 78: 7634–7638

Mohammadi M, McMahon G, Sun L, Tang C, Hirth P, Yeh BK, Hubbard SR, Schlessinger J (1997) Structures of the tyrosine kinase domain of fibroblast growth factor receptor in complex with inhibitors. Science 276: 955–960

Munoz-Sanjuan I, Brivanlou AH (2002) Neural induction, the default model and embryonic stem cells. Nat Rev Neurosci 3: 271–280

Nagy A, Rossant J, Nagy R, Abramow-Newerly W, Roder JC (1993) Derivation of completely cell culture-derived mice from early-passage embryonic stem cells. Proc Natl Acad Sci USA 90: 8424–8428

Okabe S, Forsberg-Nilsson K, Spiro AC, Segal M, McKay RDG (1996) Development of neuronal precursor cells and functional postmitotic neurons from embryonic stem cells in vitro. Mech Dev 59: 89–102

Pera EM, Wessely O, Li SY, De Robertis EM (2001) Neural and head induction by insulin-like growth factor signals. Dev Cell 1: 655–665

Pevny LH, Lovell-Badge R (1997) Sox genes find their feet. Curr Opin Genet Dev 7: 338–344

Pevny LH, Sockanathan S, Placzek M, Lovell-Badge R (1998) A role for SOX1 in neural determination. Development 125: 1967–1978

Rathjen PD, Nichols J, Toth S, Edwards DR, Heath JK, Smith AG (1990) Developmentally programmed induction of differentiation inhibiting activity and the control of stem cell populations. Genes Dev 4:2308–2318

Smith A (1998) Cell therapy: in search of pluripotency. Curr Biol 8: 802–804

Smith A (2001) Embryonic stem sells. In: Marshak, DR, Gardner RL, Gottlieb D (eds) Stem cell biology. New York, Cold Spring Harbor Laboratory Press, pp. 205–230

Smith AG (1991) Culture and differentiation of embryonic stem cells. J Tiss Cult Meth 13: 89–94

Smith AG (2001) Embryo-derived stem cells: of mice and men. Ann Rev Cell Dev Biol 17: 435–462

Solter D, Gearhart J (1999) Putting stem cells to work. Science 283: 1468-1470

Stavridis MP, Smith AG (2003) Neural differentiation of mouse embryonic stem cells. Biochem Soc Trans 31: 45–49

Streit A, Berliner AJ, Papanayotou C, Sirulnik A, Stern CD (2000) Initiation of neural induction by FGF signalling before gastrulation. Nature 406: 74–78

Thomson JA, Itskovitz-Eldor J, Shapiro SS, Waknitz MA, Swiergiel JJ, Marshall VS, Jones JM (1998) Embryonic stem cell lines derived from human blastocysts. Science 282: 1145–1147

Tropepe V, Hitoshi S, Sirard C, Mak TW, Rossant J, van der Kooy D (2001) Direct neural fate specification from embryonic stem cells: a primitive mammalian neural stem cell stage acquired through a default mechanism. Neuron 30: 65–78

Wichterle H, Lieberam I, Porter JA, Jessell TM (2002) Directed differentiation of embryonic stem cells into motor neurons. Cell 110: 385–397

Wilder PJ, Kelly D, Brigman K, Peterson CL, Nowling T, Gao Q-S, McComb RD, Capecchi MR, Rizzino A (1997) Inactivation of the FGF-4 gene in embryonic stem cells alters the growth and/ or the survival of their early differentiated progeny. Dev Biol 192: 614–629

Wilson SI, Edlund T (2001) Neural induction: toward a unifying mechanism. Nature Neurosci 4 Suppl, 1161–1168

Wilson SI, Graziano E, Harland R, Jessell TM, Edlund T (2000) An early requirement for FGF signalling in the acquisition of neural cell fate in the chick embryo. Curr Biol 10: 421–429

Wood HB, Episkopou V (1999) Comparative expression of the mouse Sox1, Sox2 and Sox3 genes from pre-gastrulation to early somite stages. Mech Dev 86: 197–201

Ye W, Shimamura K, Rubenstein JL, Hynes MA, Rosenthal A (1998) FGF and Shh signals control dopaminergic and serotonergic cell fate in the anterior neural plate. Cell 93: 755–766

Ying Q-L, Smith AG (2003a) Defined conditions for neural commitment and differentiation. In: Wassarman P, Keller G (eds) Differentiation of embryonic stem cells. Elsevier, Amsterdam

Ying Q-L, Stavridis M, Griffiths D, Li M, Smith A (2003b) Conversion of embryonic stem cells to neuroectodermal precursors in adherent monoculture. Nature Biotech 21: 183–186

Stem Cell Infidelity

J. Frisén[1]

Summary

Cells differentiate according to stereotype pedigrees, or at least so we thought. Several studies have challenged this dogma, suggesting that stem cells in several tissues may be plastic and switch lineages, but many of the results are open to other interpretations. Is there solid evidence for stem cell plasticity and should we rewrite the textbooks just yet?

Distinct lineages emerge from pluripotent cells during the succession of early embryogenesis, and progressively more restricted cells give rise to the specialized cells of different organs and tissues. Decades of developmental studies have provided us with a family tree for the generation of the major classes of cells in the body, which has unveiled robust, stereotype pedigrees. It has been thought that cells only progress in one direction along these differentiation pathways and are unable to switch tracks. In many tissues, self-renewing multipotent stem cells are maintained in adulthood and serve either to replace cells that have a limited life span or to regenerate cells after injury. Such stem cells were believed to be restricted in their potential and limited to generating the types of cells present in the tissue in which the stem cells resided. For example, a neural stem cell would be restricted to generating neural cells and an epidermal stem cell to making skin cells.

A flurry of studies over the last few years has challenged this concept, suggesting that certain tissue stem cells in embryos and adults may be more plastic than previously thought and may give rise to cells of unrelated lineages if transferred to another environment. In this new environment the stem cell would be able to respond to the novel instructive cues, which would reprogram the cell to generate cells appropriate for the new environment. This concept is known as stem cell plasticity.

However, several recent studies have, suggested alternative explanations for some of the findings ascribed to stem cell plasticity, and they have questioned the existence of this event. When studying these processes, there are numerous

[1] Department of Cell and Molecular Biology, Medical Nobel Institute, Karolinska Institute, SE-171 77 Stockholm, Sweden. Phone +46 8 728 7562, Fax +46 8 324927, Mail: jonas.frisen@cmb.ki.se

Gage et al.
Stem Cells in the Nervous System:
Functional and Clinical Implications
© Springer-Verlag Berlin Heidelberg 2004

caveats that pose a serious risk of erroneously interpreting these findings as signs of stem cell plasticity.

It is important to unravel the potential extent and molecular biology of stem cell plasticity for several reasons. First, this concept challenges our view of how cellular differentiation is regulated. Second, it poses the question whether this process may be in effect during normal physiological conditions and in pathological situations. Third, stem cells may offer an attractive source of cells for transplantation, and the concept of stem cell plasticity implies that certain adult stem cells may be much more potent and versatile than previously thought. Ultimately, they could offer an ethically uncontroversial and autologous alternative to embryonic stem cells in therapeutic situations.

Establishment and maintenance of cellular determination

To understand the potential lineage plasticity of cells, it is important to understand how cellular determination is regulated. The successful cloning of mammals demonstrates that the restriction of a cell's differentiation potential is not a result of irreversible genetic modification but rather is imposed by epigenetic constraints (Rideout et al. 2001). Preventing uniform access of the transcriptional machinery to all of the promoters in the genome is a crucial mechanism by which a cell's fate is gradually restricted during development. This may be accomplished by the use of heritable genetic imprinting mechanisms involving DNA methylation and acetyl modifications of histones, which results in the binding of specific repressors to the DNA.

Transcriptional repression of specific genes may be as important as transcriptional activation in regulating cellular determination and differentiation. Some molecular insight into this process has been gained from the study of Pax5 mutant mice. In Pax5 null mice, B lymphocyte differentiation is abolished. This transcription factor proves to be required not only for activating the expression of B cell-specific genes but also for repressing the expression of genes associated with other lineages (Busslinger et al. 2000). Even more striking is the recent evidence that the patterning of the developing nervous system and the establishment of distinct neuronal cell types are governed by transcriptional homeodomain protein repressors (Muhr et al. 2001). Most of these proteins mediate repression by interaction with the co-repressor groucho, and the identity of distinct cell types is provided by the combinatorial expression code of repressor proteins expressed in a particular cell (Muhr et al. 2001).

Some of these transcriptional programs may restrict differentiation potential by regulating a cell's sensitivity to specific instructive differentiation signals. For example, hematopoietic stem cells express certain cytokine receptors that continue to be expressed in the progenitor cells for some, but not all, lineages. If such a receptor is ectopically expressed in a progenitor, which usually does not express it, some cells can switch lineage and behave as a progenitor cell for another set of hematopoietic cells (Kondo et al. 2000). Neural stem cells, both in the central and peripheral nervous systems, initially generate mainly neurons and, later in development, mostly glial cells. During this time, neural crest stem cells, which give rise to the peripheral nervous system, switch their sensitivity from molecules that are known to instructively promote neuronal differentiation to increase their responsiveness to molecules that stimulate glial differentiation (White et al. 2001). Thus, neural crest stem cells down-regulate the expression of receptors for bone morphogenetic protein (BMP) 2, which instructively promotes neuronal differentiation, and increase the sensitivity to Notch1 signaling by increasing the ratio of Notch1 to its antagonist Numb, thereby promoting glial differentiation (Kubu et al. 2002). However, neural crest stem cells have the potential to generate neurons and glial cells both early and late in development, suggesting that rather than being strictly determined to generate a certain repertoire of cells, they have a developmentally regulated bias.

The concept of stem cell plasticity implies that cellular determination could be regulated by extracellular factors. Many stem cell properties can indeed be regulated by the local environment, or the niche, such as proliferation rate and symmetric versus asymmetric cell division. It remains to be established if a certain niche can reprogram the determination of a stem cell. Restricting a cell's differentiation potential appears to be an active process, and the need to limit the potential may decrease during development since some stem cell populations will be isolated in a niche and not exposed to signals instructing them to generate other cell types under physiological situations.

Lineage infidelity – from fruit fly to man

A burst of papers over the last few years has suggested stem cell plasticity in different experimental situations in mice. The common theme in these studies has been to follow the fate of genetically marked cells in a new environment and analyze whether cells of other lineages are generated. Among the more spectacular claims are reconstitution of hematopoiesis in irradiated mice by intraveneous injection of neural stem cells, contribution of neural stem cells to tissues of all three germ layers when injected into early chick or mouse embryos, and generation of, for example, liver cells, myocytes and even neurons from bone marrow cells (Ferrari et al. 1998; Bjornson et al. 1999; Clarke et al. 2000; Lagasse et al. 2000; Brazelton et al. 2000; Mezey et al. 2000; Krause et al. 2001). Although most of the recent studies suggesting stem cell plasticity have been performed in mice, there are examples suggesting similar phenomena in both Drosophila and man.

In Drosophila, specialized appendages such as legs and wings are formed from clusters of undifferentiated cells called imaginal discs. Cells in different imaginal discs acquire positional identity at the larval stage, long before cellular differentiation starts at metamorphosis, ensuring that appropriate appendages are formed in each location. Transplantation of cells between imaginal discs demonstrated that these cells are determined to form the appendage that is appropriate for the location the cells were taken from, and will form an inappropriate appendage for the new location to which they were transplanted. However, although most transplanted cells retain their positional identity, some transplanted cells will acquire the positional identity of the new location, a phenomenon known as transdetermination, and will form an appropriate structure for the new position (Maves and Schubinger 1999). This event is not due to loss of genetic material and inability to form cells of the original tissue since the cell, when transplanted back, may switch again to generating cells appropriate for the original location.

There are several findings indicative of stem cell plasticity in humans. By tracking cells of male origin carrying a Y chromosome in patients who have received transplants from a donor of the opposite sex, it has been possible to track the progeny of grafted cells. Several studies have demonstrated XY liver cells in women receiving transplants of male hematopoietic stem cells or bone marrow, and XY cells in the livers of males who received a liver from a female donor, suggesting that blood- and bone marrow-derived stem cells can generate hepatocytes in humans (Alison et al. 2000; Theise et al. 2000).

Multiple paths to a new identity

One can envision three conceptually different ways for a cell to switch lineages, as outlined in Figure 1. Transdifferentition is the situation where a fully differentiated cell takes on another differentiated phenotype, often without cell division. One of the best-studied situations of transdifferentiation is the generation of lens cells from retinal pigment epithelium in response to injury in newts and birds. There is also one well-described example of transdifferentiation during normal mammalian development, namely the transdifferentiation of smooth muscle cells to skeletal myocytes in the esophagus (Patapoutian et al. 1995). This process is dependent on Myf5, and in the absence of this gene there is no generation of skeletal muscle in the esophagus (Kablar et al. 2000).

The lineage switch could alternatively be executed by transdetermination. Here, a stem or progenitor cell that is determined to generate a specific set of cell types switches properties to that of another stem or progenitor cell that is determined to generate another set of differentiated cells. The best-studied example of this event is the altered determination of progenitor cells for appendages when transplanted to an imaginal disc for another appendage in Drosophila (Maves and Schubinger 1999).

The third way that a cell can switch lineage is by first dedifferentiating to a common stem or progenitor cell and then redifferentiating to another, distinct cell type. Dedifferentiation is a key step in regeneration in several amphibians,

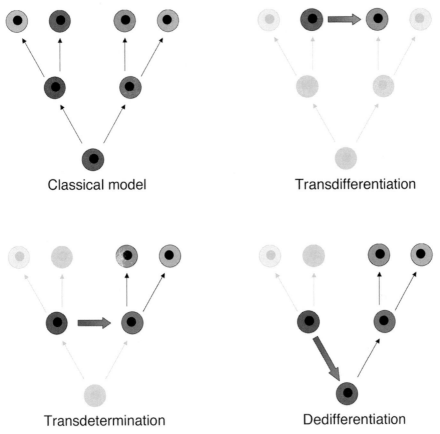

Fig. 1. How cells switch lineages. In the classical model, cells differentiate by strict progression along different branches of a linear pedigree. In this image, a common stem cell (at the bottom of the family tree) generates two stem cells that are determined to generate two different specialized cell types. Three conceptually different ways in which a cell can switch lineage are depicted. A differentiated cell can take on the phenotype of another differentiated cell, known as transdifferentiation. An example of this is the generation of a lens from retinal pigment cells after eye injury in newts. Transdetermination is the situation in which a stem cell that is determined to generate cells of a certain lineage switches to another stem cell state and generates progeny of the latter lineage. This can be seen when cells are transplanted between imaginal discs in Drosophila larvae. Dedifferentiation to a common, more potent stem cell, followed by differentiation along another lineage, is an alternative way for a cell to switch lineage. The cell that dedifferentiates could be a fully differentiated cell, or as in this image, a determined stem cell. Dedifferentiation is seen, for example, after limb amputation in newts, which results in dedifferentiation of local myocytes, followed by regeneration of cells of different lineages.

where the newt has served as a valuable model organism. In response to limb amputation in newts, differentiated cells at the site of the injury are induced to dedifferentiate and proliferate in response to local extracellular factors.

Proliferation of dedifferentiated cells gives rise to a regeneration blastema from which new differentiated cells are formed that build up the new limb (Brockes 1997). Some of the molecular mechanisms for dedifferentiation appear to be conserved, since some mammalian cells can be induced to dedifferentiate in response to newt blastema extract (McGann et al. 2001). Moreover, Msx1, a transcription factor expressed during limb development and re-expressed during limb regeneration in amphibians, induces multinucleated myotubes to dedifferentiate and cleave to form dividing mononucleated cells when ectopically expressed in a mammalian myocyte cell line. More importantly, these cells were no longer restricted to generating cells belonging to the myocyte lineage but could now also generate adipocytes, chondrocytes and osteoblasts (Odelberg et al. 2000).

It is not trivial to distinguish between transdifferentiation, transdetermination and dedifferentiation followed by redifferentiation. It requires following the expression of markers or functional properties of cells at different stages of the process. An inherent problem is that, if you kill a cell for analysis of markers, even if you knew the identity of the cell at the onset of the experiment, you will never know what it was going to become. This problem can be circumvented by the use of markers detectable in live cells, for example, fluorescent reporters expressed under the control of specific promoters. Another limitation is that markers for different cells in the potential pathways have to be well characterized. Importantly, in no study suggesting stem cell plasticity in mammals have data been provided indicating in which of the above-outlined ways a certain cell may have switched lineage.

An alternative explanation to what may appear as lineage switching has been put forward. In this model, all tissue stem cells are the same and are instructed by their environment to generate specific progeny appropriate for their location (Blau et al. 2001). What argues against this model is that different tissue-specific stem cells do have distinct marker profiles and functional features and are, if not absolutely committed, at least strongly biased to form certain cell types even when brought out of their normal context.

Phenomenology, pathology or physiology?

It is important to underscore that many of the experimental situations where stem cell plasticity has been implied are artificial and may be far from reflecting a physiological situation. Some may resemble situations of pathology, whereas others do not resemble situations seen in real life but are purely experimental, such as for example injecting adult cells into embryos. Moreover, in some experiments stem cells are cultured before being exposed to a new environment, which may make them lose their positional bearings and be more amenable to switching lineage. Therefore, it is important, as phrased in a recent review, to distinguish between the actual and the possible (Anderson 2001). Although some of the experimental situations are far from what is seen in physiological situations, it may be informative to study how, for example, cellular differentiation and determination are influenced in extreme situations. This is underscored in recent studies of cellular reprogramming by nuclear transfer, an artificial situation that promises to teach us much about fundamental aspects of cellular determination (Rideout et al. 2001).

In many situations where stem cell plasticity has been implied there has been damage to the tissue where the new cells of unrelated lineage emerge, and there are some indications that this damage may be a prerequisite for the observed effects. For example, extremely few hematopoietic stem cell-derived liver cells are seen in the uncompromised liver, whereas pronounced contribution has been reported in pathological situations (Lagasse et al. 2000). This finding could imply that stem cell plasticity is a rare phenomenon and may reach appreciable levels only if these cells have a competitive advantage.

Another situation of human pathology where stem cell plasticity may be implied is in metaplasia, i.e., when there is a change of cell type in a certain location as an effect secondary to a pathological process (Slack and Tosh 2001). Common examples of metaplasia include pancreas cells in the intestine, gastric epithelium in the duodenum and endometrium in the ovary. Metaplasias are thought to be polyclonal and provide a way for local stem or progenitor cells to adapt to a changed environment by producing cells appropriate for the new condition.

Is there any evidence for stem cell plasticity during normal development? Although most cells are thought to progress in their differentiation along stereotype pedigrees, nature offers some quirks. For example, ectodermal neural crest cells give rise to what we in other situations consider mesodermal derivatives such as muscle, connective tissue, cartilage and bone. Another example is epidermal placodes, thickenings of the primitive skin, which by induction from underlying structures form neural tissue. In both neural crest and placode cells, the generation of progeny that are not classically lineage-related is instructed by the environment. Both are examples that may be viewed as switching lineages – in the case of neural crest from ectodermal to mesodermal, and in the case of placodes from epidermal to neural, – and may be interpreted as stem cell plasticity.

Fact or fusion?

Cellular fusion results in a situation similar to that created in nuclear transfer experiments, where the nucleus of a cell fusing with another will be influenced by the epigenetic signals in the cytoplasm of the new partner. This fusion can result in reprogramming to a specific lineage or a pluripotent state, depending on the cell type with which the studied cell fused.

Two recent studies suggested that many of the results interpreted as stem cell plasticity may be due to cell fusion (Terada et al. 2002; Ying et al. 2002). They found that in co-cultures of embryonic stem cells with brain or bone marrow cells, pluripotent hybrid cells emerged spontaneously. In neither study were stem cells from brain or bone marrow, respectively, required for this event, since the frequency was the same when comparing unfractionated brain cells with cultured neural stem cells or whole bone marrow with purified hematopoietic stem cells. Although the frequency of this event was extremely low (1:10,000–100,000 brain cells or 1:100,000–1,000,000 bone marrow cells), it stresses that adult somatic cells can gain differentiation potential by fusion with less differentiated cells (Terada et al. 2002; Ying et al. 2002), and the frequency could be higher in other situations. Both of these studies analyzed systems conceptually different from those in which stem cell plasticity has been suggested, therefore precluding conclusions to be drawn regarding the previous studies. However, they do point out an important caveat in these types of experiments which has been overlooked and needs to be addressed in future studies.

Some results that have been perceived as stem cell plasticity appear very unlikely to be due to fusion. For example, the potential of bone marrow cells to generate liver cells is very robust in certain models (Lagasse et al. 2000). Moreover, all capacity in the bone marrow to form hepatic cells is found within the hematopoietic stem cell population (Lagasse et al. 2000), demonstrating that it requires stem cells, in contrast to the fact that the emergence of pluripotent cells in co-cultures of bone marrow with embryonic stem cells is independent of hematopoietic stem cells (Terada et al. 2002; Ying et al. 2002). Another example of suggested stem cell plasticity that is unlikely to require fusion is the generation of myocytes from neural stem cells. This event has been described in several systems, but most importantly in this context, it has been shown to occur without exposure to heterologous cells. Thus, neural stem cells differentiate to cells expressing myocyte markers and display the morphology of muscle cells when cultured at very low density or when exposed to BMPs (Tsai and McKay 2000).

On the other hand, it is likely that some examples of suggested stem cell plasticity involve fusion between the progeny of grafted stem cells and cells of the host. Several cell types are normally bi- or multinucleated, and recent experiments have proposed that some of these cell types, such as myocytes, hepatocytes and Purkinje neurons, have been generated from bone marrow-derived stem cells. When taking the possibility of fusion into account, an alternative explanation to the observed phenomena could then be that a transplanted cell, or its progeny, has fused with a preexisting cell, rather than formed such a cell de novo. However, fusion is the way certain tissues like muscle develop normally. The fact that bone marrow-derived cells fuse with multinucleated myofibers does not necessarily mean that this is not an example of stem cell plasticity. Detailed studies of whether bone marrow-derived cells fuse directly with myofibers or whether they enter myocyte differentiation to become a myoblast prior to fusion will provide further insights regarding the true nature of this phenomenon.

Fusion appears to be an extremely rare event and may seem an unlikely explanation for many of the results ascribed to stem cell plasticity. On the other hand, based on previous knowledge, stem cell plasticity also seems unlikely. For each individual situation where a suspected case of plasticity is seen, it will have to be established which of the two unlikely possibilities, is true: fusion or plasticity. At this point it is as unwise to conclude that tissue stem cells are completely plastic as to conclude that all data suggesting stem cell plasticity are due to fusion events. We need to keep our minds both open and very skeptical.

How can stem cell plasticity be tested?

To comprehend the possible extent of stem cell plasticity and to be able to understand the molecular underpinning, it is important to be rigorous when defining it, not least since there are ample caveats and opportunities to misinterpret findings as being indicative of stem cell plasticity (Fig. 2).

A first key issue is to define the cell that is studied. In many cases heterogeneous groups of cells have been studied, which precludes conclusions regarding lineage conversion. Unless starting with a homogeneous population of cells (ideally a single, prospectively identified cell), it is difficult to exclude that the starting population contained a mix of stem or progenitor cells for different lineages. For example, many studies have described the generation of various non-hematopoietic cell types, such as liver and muscle cells, from transplanted bone marrow. Bone marrow is, however, a complex tissue that harbors multiple different cell types and lineages, and experiments showing myocytes or hepatocytes being derived from unfractionated bone marrow could in theory equally well be interpreted as showing the presence of muscle and liver progenitor cells in bone marrow as plasticity of hematopoietic stem cells residing in bone marrow.

A related caveat may be the presence of a small number of heterologous stem cells in certain tissues. For example, cells isolated from muscle based on dye exclusion properties common to several stem cells have hematopoietic potential, and was initially interpreted as a case of stem cell plasticity where muscle stem cells could generate blood. More recently, characterization of the isolated muscle cells has suggested that the hematopoietic potential in muscle resides exclusively in a population of cells with the marker profile of hematopoietic stem cells, suggesting that the presence of such cells in circulation may explain this result (McKinney-Freeman et al. 2002).

Yet another source of misinterpretation could be that transformed cells may generate unrelated cell lineages. The reconstitution of hematopoiesis in irradiated mice by intraveneous injection of neural stem cells (Bjornson et al.

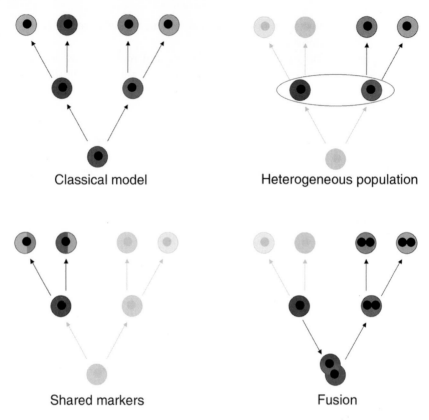

Fig. 2. Situations that may be misinterpreted as stem cell plasticity. Situations that represent differentiation of cells along a linear pedigree may easily be interpreted as suggesting stem cell plasticity. It is crucial to be certain of the starting point of the analysis, i.e., the cell type whose progeny is analyzed. Many tissues may contain heterogeneous cell populations. For example, hematopoietic stem cells are present in the circulation and in many tissues, and hematopoietic potential of cells from a certain tissue may be the result of either the plasticity of a non-hematopoietic stem cell or the presence of hematopoietic stem cells in that tissue. The analyzed cell must be characterized either by phenotype – which is well established based on cell surface markers in, for example, the hematopoietic system – or by function, which is readily done in vitro in, for example, the case of neural stem cells. There are few cellular markers that are truly specific for a certain cell type, and it is thus dangerous to conclude stem cell plasticity based on the expression of a few markers. Moreover, a cell may phagocytize a cell of another lineage and in that way, at least transiently, acquire molecular markers of a different lineage. The unreliability of molecular markers in firmly establishing the identity of a certain cell underscores the need for evidence of function appropriate for the particular cell type. This is in many cases best shown by experiments where a certain cell type can rescue the phenotype of a mutant lacking a certain cell type or function. Finally, cell fusion may result in hybrid tetraploid cells that will carry genetic markers of both cells. A cell that fuses with another cell may be reprogrammed by factors in the cytoplasm of the new partner and take on properties of this cell. For example, fusion of neural and bone marrow cells with embryonic stem cells occurs spontaneously at a very low frequency in vitro and results in pluripotent cells that can generate a variety of cell types to which these somatic cells do not normally give rise.

1999), was recently suggested to be a result of transformation of neural stem cells by excessive passaging in vitro (Morshead et al. 2002).

Equally important as defining the starting point for potential plasticity is characterizing the end point, i.e., the cell type generated from the stem cell of interest. The identity of a cell is often defined by morphology and expression of appropriate markers. Ideally, one should demonstrate function of the generated cell, although this may be very difficult in many cases. This demonstration was done elegantly in one study to date, in which prospectively identified hematopoietic stem cells were found not only to generate liver cells with a marker profile expected by hepatocytes but also to rescue mice with a genetic defect resulting in lethal liver failure (Lagasse et al. 2000). Although establishing the function of cells generated by potential stem cell plasticity will be very important and eventually necessary, it is probably overzealous and counterproductive to demand establishment of function in each individual study, since this is excruciatingly difficult in some cases. For example, although adult neurogenesis has been intensively studied for the last decade, it was only recently established that adult-born neurons are functional (Carlén et al. 2002; van Praag et al. 2002).

Finally, as discussed above, cellular fusion may be misinterpreted as stem cell plasticity. Two cells that have fused to become tetraploid can be distinguished by karyotyping. In tissue sections one can use FISH to analyze the number of sex chromosomes, which will reveal four instead of two, or by flowcytometry analysis of the DNA complement of cells in solution. However, tetraploid cells may expel supernumerary chromosomes and approach a diploid DNA complement. Analysis of distinct genetic markers is therefore required to conclusively establish whether a certain cell has a mixed karyotype with chromosomes from two different cells.

Conclusion

Transfer of nuclei of somatic cells to oocytes and the cloning of adult animals have in a striking way illustrated that, in most cells, there are no irreversible changes to the genome as cells differentiate, but rather that the differentiated state is established and maintained by epigenetic signals. The concept of stem cell plasticity implies that not only may cells be reprogrammed in such extreme situations in which the intracellular milieu is switched by nuclear transfer or cell fusion but also that extracellular signals can reprogram cells to switch lineages. However, the fact is that none of the studies suggesting plasticity of adult stem cells has excluded all alternative explanations. Thus, it may be wise to await further studies before we revise our view of how cellular differentiation is regulated and we rewrite the textbooks.

Acknowledgments

Work in the author's laboratory was supported by the Swedish Foundation for Strategic Research, the Swedish Research Council, the Karolinska Institute and the Swedish Cancer Foundation.
Figure legends

References

Alison MR, Poulsom R, Jeffery R, Dhillon AP, Quaglia A. Jacob J, Novelli M, Prentice G, Williamson J, Wright NA (2000) Hepatocytes from non-hepatic adult stem cells. Nature 406,: 257.

Anderson DJ (2001) Stem cells and pattern formation in the nervous system: the possible versus the actual. Neuron 30: 19–35.

Bjornson CR, Rietze RL, Reynolds A, Magli MC, Vescovi AL (1999) Turning brain into blood: a hematopoietic fate adopted by adult neural stem cells in vivo. Science 283: 534–537.

Blau H, Brazelton TR, Weimann JM (2001) The evolving concept of a stem cell: entity or function. Cell 105: 829–841.

Brazelton TR, Rossi FM, Keshet GI, Blau HM (2000) From marrow to brain: expression of neuronal phenotypes in adult mice. Science 290: 1775–1779.

Brockes JP (1997) Amphibian limb regeneration: rebuilding a complex structure. Science 276: 81–87.

Busslinger M, Nutt SL, Rolink, AG (2000) Lineage commitment in lymphopoiesis. Curr Opin Immunol 12,: 151–158.

Carlén M, Cassidy RM, Brismar H, Smith GA, Enquist LW, Frisén J (2002) Functional integration of adult-born neurons. Curr Biol 12: 606–608.

Clarke DL, Johansson CB, Wilbertz J, Veress B, Nilsson E, Karlström H, Lendahl U, Frisén J (2000) Generalized potential of adult neural stem cells. Science 288: 1660–1663.

Ferrari G, Cusella-De Angelis G, Coletta M, Paolucci E, Stornaiuolo A, Cossu G, Mavilio F (1998) Muscle regeneration by bone marrow-derived myogenic progenitors. Science 279,: 1528–1530.

Kablar B, Tajbakhsh S, Rudnicki MA (2000) Transdifferentiation of esophageal smooth to skeletal muscle is myogenic bHLH factor-dependent. Development 127: 1627–1639.

Kondo M, Scherer DC, Miyamoto T, King AG, Akashi K, Sugamura K, Weissman IL (2000) Cell-fate conversion of lymphoid-committed progenitors by instructive actions of cytokines. Nature 407: 383–386.

Krause DS, Theise ND, Collector MI, Henegariu O, Hwang S, Gardner R, Neutzel S, Sharkis SJ (2001) Multi-organ, multi-lineage engraftment by a single bone marrow-derived stem cell. Cell 105: 369–377.

Kubu CJ, Orimoto K, Morrison SJ, Weinmaster G, Anderson DJ, Verdi JM (2002) Developmental changes in Notch1 and Numb expression mediated by local cell-cell interactions underlie progressively increasing delta sensitivity in neural crest stem cells. Dev Biol 244: 199–214.

Lagasse E, Connors H, Al-Dhalimy M, Reitsma M, Dohse M, Osborne L, Wang X, Finegold M, Weissman IL, Grompe M (2000) Purified hematopoietic stem cells can differentiate into hepatocytes in vivo. Nature Med 6: 1229–1234.

Maves L, Schubinger G (1999) Cell determination and transdetermination in Drosophila imaginal discs. Curr Top Dev Biol 43:115–151.

McGann CJ, Odelberg SJ, Keating MT (2001) Mammalian myotube dedifferentiation induced by newt regeneration extract. Proc Natl. Acad Sci USA 98: 13699–13704.

McKinney-Freeman SL, Jackson KA, Camargo FD, Ferrari G, Mavilio F, Goodell MA (2002) Muscle-derived hematopoietic stem cells are hematopoietic in origin. Proc Natl Acad Sci USA 99:1341–1346.

Mezey E, Chandross KJ, Harta G, Maki RA, McKercher SR (2000) Turning blood into brain: cells bearing neuronal antigens generated in vivo from bone marrow. Science 290: 1779–1782.

Morshead CM, Benveniste P, Iscove NN, van der Kooy D (2002) Hematopoietic competence is a rare property of neural stem cells that may depend on genetic and epigenetic alterations. Nature Med 8: 268–273.

Muhr J, Andersson E, Persson M, Jessell TM, Ericson J (2001) Groucho-mediated transcriptional repression establishes progenitor cell pattern and neuronal fate in the ventral neural tube. Cell 104: 861–873.

Odelberg SJ, Kollhof A, Keating MT (2000) Dedifferentiation of mammalian myotubes induced by msx1. Cell 103: 1099–1109.

Patapoutian A, Wold BJ, Wagner RA (1995) Evidence for developmentally programmed trans-differentiation in mouse esophageal muscle. Science 270: 1818–1821.

Rideout WM, Eggan K, Jaenisch R (2001) Nuclear cloning and epigenitic reprogramming of the genome. Science 10: 1093–1098.

Slack JMW, Tosh D (2001) Transdifferentiation and metaplasia – switching cell types. Curr Opin Gen Dev 11: 581–586.

Terada N, Hamazaki T, Oka M, Hoki M, Mastalerz DM, Nakano Y, Meyer EM, Morel L, Petersen BE, Scott EW (2002) Bone marrow cells adopt the phenotype of other cells by spontaneous cell fusion. Nature 416: 542–545.

Theise ND, Nimmakayalu M, Gardner R, Illei PB, Morgan G, Teperman L, Henegariu O, Krause DS (2000) Liver from bone marrow in humans. Hepatology 32: 11–16.

Tsai RYL, McKay RDG (2000) Cell contact regulates fate choice by cortical stem cells. J. Neurosci. 20: 33725–3735.

van Praag H, Schinder AF, Christie BR, Toni N TD, Palmer TD, Gage FH (2002) Functional neurogenesis in the adult hippocampus. Nature 415: 1030–1034.

White PM, Morrison SJ, Orimoto K, Kubu CJ, Verdi JM, Anderson DJ (2001) Neural crest stem cells undergo cell-intrinsic developmental changes in sensitivity to instructive differentiation signals. Neuron 29: 57–71.

Ying QL, Nichols J, Evans EP, Smith AG (2002) Changing potency by spontaneous fusion. Nature 416: 545–548.

The Importance of CNS Stem Cells in Development and Disease

R. Mc Kay[1]

Summary

Work from our group played an important role in the early period when stem cells of the central nervous system were first defined (McKay 1997). Our recent work contributes to understanding three fundamental processes in the developing nervous system: 1) cell cycle control (Tsai and McKay 2002), 2) the control of cell fate (Panchision et al. 2001) and 3) the early steps in neuronal differentiation (Blondel et al. 2000; Vicario-Abejon et al. 2001; Collin et al. 2001). Work on CNS stem cells has developed to a stage where there are also clinical implications. Recent advances in the application of stem cell biology to Parkinson's disease clearly demonstrate the potential importance of a strong scientific foundation in the use of stem cells in models of neuronal loss or injury (Studer et al 1998, 2000; Lee et al. 2000; Sanchez-Pernaute et al. 2001; Kim et al. 2002). The potential of these methods is demonstrated by our contributions in other areas, for example in glial transplantation (Brustle et al. 1999), adult neurogenesis (Cameron and McKay 1999, 2001) and endocrine pancreatic differentiation Lumelsky et al. 2001). We study both stem cells and the early functions of neurons because we believe that the functional analysis of differentiating neurons is the best measure of the potential of a stem cell. It is also our view that new knowledge of both stem cell biology and neuronal differentiation is essential to fulfill the clinical potential of this field.

It is many years since evidence was first obtained for multipotential precursor cells in the vertebrate and invertebrate nervous systems (Nieuwkoop 1952; Ready et al. 1976; Le Douarin 1980). In our group, in vivo studies were first used to identify CNS stem cells (Hockfield and McKay 1985; Lendahl, 1990). The identification of the cell was rapidly followed by evidence for the developmental plasticity of CNS stem cells obtained by transplantation into ectopic sites in the CNS (Frederiksen et al. 1988; Renfranz et al. 1991; Vicario-Abejon et al. 1995). The potential of CNS stem cells expanded in vitro was also defined (Temple 1989; Davis and Temple 1994; Cattaneo and McKay 1990; Reynolds and Weiss

[1] Lab. of Molecular Biology, NIDS, NIH, Bldg 36 RM 5A29, Convent Dr-MSC 4157, BETHESDA, MD 20892-4157, USA, Phone 301 496 6574, Télécopie 301 402 4738, Mail: mckay@codon.nih.gov

Gage et al.
Stem Cells in the Nervous System:
Functional and Clinical Implications
© Springer-Verlag Berlin Heidelberg 2004

1992). The molecular mechanisms controlling CNS stem cell numbers and fate choice are now the focus of work in many laboratories using both in vivo and in vitro systems (Isaka et al. 1999; Mitsuhashi et al. 2001; Sun et al. 2001; Petersen et al. 2002; Hermanso et al. 2002; Turnley et al. 2002).

A central goal of our work is to understand the molecular mechanisms controlling the switch from multipotency to commitment. During development, precursor cells in the nervous system assume a positional identity within a spatial coordinate system that has a dorsal-ventral, rostral-caudal and left-right axis(Altmann and Brivanlou 2001). When these cells subsequently undergo mitotic arrest, they rapidly acquire characteristics of a specific terminal fate. The differentiation of stem cells in vitro suggests that many of these specific features occur in the absence of the precise organization that occurs in vivo. The self-organization shown during stem cell differentiation suggests that a set of concerted responses can either maintain the precursor state or direct the cell to a set of related fates. We used two different techniques to study the molecular basis of stem cell commitment. In the first set of experiments, we studied the action of the morphogen BMP on stem cells in vitro and in vivo. These data suggest that BMPs function through distinct receptors to achieve coordinated effects on stem cell proliferation, identity and terminal differentiation (Panchision et al. 2001). In the second set of experiments we identified a new gene that is expressed in stem cells in different tissues that may be involved in a switch in the regulation of p53 that occurs when proliferating stem cells transit into committed progenitor cells (Tsai and McKay 2002). p53 regulates apoptosis and is mutated in more than half of human cancers (Bullock et al. 2000). We propose experiments to extend our understanding of the function of BMP signaling and p53 regulators in the commitment of a CNS stem cell to a terminally differentiated state. The stem cell-specific regulation of p53 may give us insight into the role of stem cells in cancer and illustrates the potential general significance of stem cell biology.

Evidence that stem cells can make functionally mature progeny has contributed importantly to the interest in stem cells. Neurons are perhaps best defined by their ability to form synapses. Neuronal differentiation is not a simple, cell-autonomous process, as interactions with astrocytes are critical to synapse formation (Blondel et al. 2000; Collin et al. 2001; Mauch et al. 2001; Mozhayeve et al. 2002). In the presence of astrocytes, hippocampal stem cells from the fetal and adult nervous system generate neurons with glutamatergic and GABA-ergic synapses (Vicario-Abejon et al. 2000; Song et al. 2002). Behavioral and electrophysiological tests demonstrate that functional dopaminergic neurons can also be derived from expanded midbrain precursors (Studer et al. 1998) and from embryonic stem (ES) cells (Lee et al. 2000; Kim et al. 2002). In addition, motor neurons (Wichterle et al. 2002) and functional glia (Brustle et al. 1999) have been obtained from ES cells. These experiments suggest that stem cells expanded in vitro can differentiate to functional progeny. We propose experiments to more accurately define the differentiation and survival mechanisms of functional neurons derived from telencephalic and mesencephalic stem cells. There is increasing evidence for the replacement of functional neurons from endogenous cells in the adult CNS. In the dentate gyrus and olfactory bulb, neurogenesis normally occurs at a significant rate (Fuchs and Gould 2000; (Cameron and McKay 2001; van Praag et al. 2002). Endogenous cells may also replace neurons in the adult cortex, striatum and hippocampus after injury (Magavi et al. 2000; Benraiss et al. 2001; Nakatomi et al. 2002; Arvidsson et al. 2002). The combination tools that are now available permit in vitro and in vivo studies to identify the mechanisms generating the major types of neurons in the developing and adult brain. Our growing understanding of neurogenesis suggests that it may be possible to stimulate neuron formation in many sites in the adult brain where neurogenesis does not normally occur at a significant rate.

Interest in stem cells has been greatly stimulated by the isolation of human ES cells and by their potential use in cell-based therapies (Thomson et al. 1998). Mouse ES cells can be induced to generate many cell types, including CNS stem cells, hematopietic stem cells, cardiomycytes, pancreatic islets and the cells of the vascular system (Lacaud et al. 2002; Rideout et al. 2002; Yamashita et al. 2000; Lumelsky et al. 2001). The major focus in our lab has been the generation of dopamine neurons, as discussed in more detail below. However, as we developed this technology. it occurred to us that the similarity in developmental mechanisms in the brain and the endocrine pancreas may make it possible to generate the cells of the endocrine pancreas and CNS with the same ES cell differentiation method. Following our initial demonstration of this possibility, two other groups also generated islet-like cells from ES cells using methods that are close to the multistage protocol initially developed in our lab (Lumelsky et al. 2001). These studies also extended this work in important ways (Hori et al. 2002; Blyszczuk et al. 2003). They both show that these ES-derived islets function in animals to regulate glucose levels.

Access to large numbers of quality-controlled cells is a critical step in developing new therapies for Parkinson's disease. Currently, fetal tissue for clinical use is hard to obtain and stem cell technology may solve this limitation. A major goal of our group is to generate large numbers of rodent and human midbrain dopamine neurons. It is our hope that access to large numbers of neurons will accelerate the development of new approaches to Parkinson's disease and encourage the use of stem cell technology more widely in neurology (Fig. 1). Our ability to approach the regeneration of the dopaminergic system is limited by our poor understanding of the induction, differentiation and integration of dopaminergic neurons in the fetal CNS. This limitation is very clear in our work on human fetal midbrain. These cells, like their rat counterparts, can be expanded in tissue culture and differentiated into dopamine neurons that function in a lesioned brain (Sanchez-Pernaute et al. 2001). However, we cannot expand either human or rodent midbrain precursors for more than a few divisions before they lose their potential to generate dopamine neurons (Yan et al. 2001). Midbrain precursors derived from ES cells also show limited

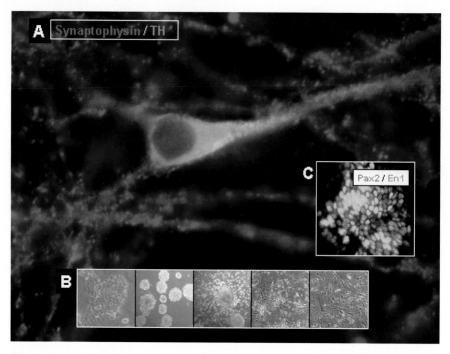

Fig. 1. Mouse ES cells generate tyrosine-hydroxylase (TH)-positive dopamine neurons (A) after a standard 5-stage differentiation procedure (B) The efficient production of midbrain precursors is shown by the co-expression of the transcription factors Pax2 and Engrailed 1 (C).

proliferation potential. One possible explanation is that the precursor cells of the ventral midbrain alter their dorsal-ventral identity as they are grown in culture (Panchision et al. 2001).

Gene disruption experiments show that the Nurr1, Lmx1b, Ptx3, En-1 and En-2 genes are expressed in, and regulate the differentiation of, dopamine precursors. Nurr1 encodes an orphan nuclear hormone receptor. Nurr1 disruption leads to the selective loss of tyrosine hydroxylase (TH)$^+$ neurons in the substantia nigra (SN) and ventral tegmental area (VTA) of the ventral mesencephalon. TH$^+$ cells appear at numbers comparable to wild type controls and possess normal morphology in all other parts of the CNS and PNS (Zetterstrom, 1997; Saucedo-Cardenas et al. 1998) Markers for dopaminergic progenitors, such as aldehyde hydrogenase 2 (AHD2) and the homeobox transcription factors, En-1 and Ptx3, are expressed in Nurr1-deficient mice, confirming that Nurr1 is not involved in the early steps of midbrain specification (Wallen, 1999). In *Nurr1*$^{-/-}$ mice, TH expression appears transiently in the developing midbrain but c-RET expression was not seen (Wallen, 2001). This gene may play an important role in the dopamine neuron as it encodes the receptor for ligands in the GDNF family that are known to have major effects on the dopamine neurons and are candidates for gene therapy approaches to Parkinson's disease. Midbrain cells from *Nurr1*[-/-] mice can express TH and are stable in vitro, suggesting that the expression of TH is not only controlled by Nurr1 [Tornqvist, 2002). We have shown that expressing high levels of the *Nurr1* gene in ES cells leads to larger numbers of dopamine neurons (Kim et al. 2002). One possible explanation for the successful outcome of Nurr1-expressing ES cells is that the introduced gene is signaling through dimerization with the RXR receptor (Aarnisalo, 2002).

The transplantation of dopamine neurons derived from over-expressing Nurr1 ES cells to the hemiparkisonian rats has shown that these cells innervate the lesioned brain and positively influence the standard behavioral tests used to evaluate this animal model. The recovery observed in drug-induced rotational behavior is attributed to the release of dopamine by the graft in the lesioned hemisphere. The grafted animals also show recovery in a battery of non-pharmacological tests, which provide a more direct measure of motor deficits analogous to limb akynesia and sensory-motor impairments in human Parkinson's disease (Kim et al. 2002). We have also shown that ES cell-derived dopamine neurons have electrophysiological properties that are expected of midbrain dopamine neurons that differentiate in vivo (Kim et al. 2002). These results encourage the in vitro generation of dopamine neurons as a tool in the study of Parkinson's disease. However, the precise properties of these cells must be more carefully defined before their exact equivalence to in vivo-generated neurons can be fully established.

Human ES cells generate neural fates in a similar way to mouse ES cells (Thomson et al. 1998; Reubinoff et al. 2000; 2001). We propose to use our understanding of the basic biology of rodent CNS stem cells to differentiate human ES cells to a specific neural fate, the midbrain dopamine neuron. The rapid progress occurring in stem cell biology suggests that it may be possible to generate human tissues by the controlled differentiation of stem cells. These results suggest that a powerful technology may develop to manipulate human ES cells and this prospect opens many clinically significant paths. Our own work has encouraged the idea that stem cell technologies might contribute to cell therapies for Parkinson's disease, multiple sclerosis and diabetes. In the future, we will focus our human ES cell work on developing robust methods to generate midbrain dopamine neurons. Our aim is to allow access to large numbers of human dopamine neurons that can be manipulated and analyzed by groups working on different aspects of Parkinson's disease. The clinical significance of stem cell biology is not restricted to cell replacement but we believe we can make a significant contribution by focusing on the specific goal of generating the nigrostriatal projection neuron from ex vivo-expanded stem cells.

References

Aarnisalo P, Kim CH, Lee JW, Perlmann T (2002) Defining requirements for heterodimerization between the retinoid X receptor and the orphan nuclear receptor Nurr1. J Boil Chem 277: 35118–35123

Altmann CR, Brivanlou AH (2001) Neural patterning in the vertebrate embryo. Int Rev Cytol 203: 447–482

Arvidsson A, Collin T, Kirik D, Kokaia Z, Lindvall O (2002) Neuronal replacement from endogenous precursors in the adult brain after stroke. Nature Med 8:963–970

Benraiss A, Chmielnicki E, Lerner K, Roh D, Goldman SA (2001) Adenoviral brain-derived neurotrophic factor induces both neostriatal and olfactory neuronal recruitment from endogenous progenitor cells in the adult forebrain. J Neurosci 21:6718–6731

Blondel O, Collin C, McCarran WJ, Zhu S, Zamostiano R, Gozes I, Brenneman DE, McKay RD (2000) A glia-derived signal regulating neuronal differentiation. J Neurosci 20: 8012–8020

Blyszczuk P, Czyz J, Kania G, Wagner M, Roll U, St-Onge L, Wobus AM (2003) Expression of Pax4 in embryonic stem cells promotes differentiation of nestin-positive progenitor and insulin-producing cells. Proc Natl Acad Sci USA 100:998–1003

Brustle O, Jones KN, Learish RD, Karram K, Choudhary K, Wiestler OD, Duncan ID, McKay RD (1999) Embryonic stem cell-derived glial precursors: a source of myelinating transplants. Science 285:754–756

Bullock AN, Henckel J, Fersht AR (2000) Quantitative analysis of residual folding and DNA binding in mutant p53 core domain: definition of mutant states for rescue in cancer therapy. Oncogene 19: 1245–5612

Cameron HA, McKay RD (1999) Restoring production of hippocampal neurons in old age. Nat.Neurosci. 2: 894–897

Cameron HA, McKay RD (2001) Adult neurogenesis produces a large pool of new granule cells in the dentate gyrus. J Comp Neurol 435: 406–417

Cattaneo E, McKay R (1990) Proliferation and differentiation of neuronal stem cells. Nature 347: 762–765

Collin Cl, Vicario-Abejon C, Rubio ME, Wenthold RJ, McKay RD, Segal M (2001) Neurotrophins act at presynaptic terminals to activate synapses among cultured hippocampal neurons. Eur J Neurosci 13: 1273–1282

Davis AA, Temple S (1994) A self-renewing multipotential stem cell in embryonic rat cerebral cortex. Nature, 372:263–266

Frederiksen K, McKay RD (1988) Proliferation and differentiation of rat neuroepithelial precursor cells in vivo. J Neurosci 8: 1144–1151

Frederiksen K, Jat PS, Valtz N, Levy D, McKay R (1988) Immortalization of precursor cells from the mammalian CNS. Neuron 1:439–448

Fuchs E, Gould E (2000) In vivo neurogenesis in the adult brain: regulation and functional implications. Eur J Neurosci 12: 2211–2214

Hermanson O, Jepsen K, Rosenfeld MG (2002) N-CoR controls differentiation of neural stem cells into astrocytes. Nature 419: 934–939

Hockfield S, McKay RD (1985) Identification of major cell classes in the developing mammalian nervous system. J Neurosci 5: 3310–3328

Isaka F, Ishibashi M, Taki W, Hashimoto N, Nakanishi S, Kageyama R (1999) Ectopic expression of the bHLH gene Math1 disturbs neural development. Eur J Neurosci 11:2582–2588

Hori Y, Rulifson IC, Tsai BC, Heit JJ, Cahoy JD, Kim SK (2002) Growth inhibitors promote differentiation of insulin-producing tissue from embryonic stem cells. Proc Natl Acad Sci USA 99:16105–16110

Kim JH, Auerbach JM, Rodriguez-Gomez JA, Velasco I, Gavin D, Lumelsky N, Lee SH, Nguyen J, Sanchez-Pernaute R, Bankiewicz K, McKay R (2002) Dopamine neurons derived from embryonic stem cells function in an animal model of Parkinson's disease. Nature 418:50–56

Lacaud G, Gore L, Kennedy M, Kouskoff V, Kingsley P, Hogan C, Carlsson L, Speck N, Palis J, Keller G (2002) Runx1 is essential for hematopoietic commitment at the hemangioblast stage of development in vitro. Blood100:458–466

Le Douarin N (1980) Migration and differentiation of neural crest cells. Curr Top Dev Biol 16: 31–85

Lee SH, Lumelsky N, Studer L, Auerbach JM, McKay RD (2000) Efficient generation of midbrain and hindbrain neurons from mouse embryonic stem cells. Nature Biotechnol. 18:675–679

Lendahl U, Zimmerman L, McKay RDG (1990) CNS stem cells express a new class of intermediate filament protein. Cell 60: 585–595

Lumelsky, N, Blondel O, Laeng P, Velasco I, Ravin R, McKay R (2001) Differentiation of embryonic stem cells to insulin-secreting structures similar to pancreatic islets. Science. 292:1389–1394

Magavi SS, Leavitt BR, Macklis JD (2000) Induction of neurogenesis in the neocortex of adult mice. Nature 405: 951–955

Mauch DH, Nagler K, Schumacher S, Goritz C, Muller EC, Otto A, Pfrieger FW (2001) CNS synaptogenesis promoted by glia-derived cholesterol. Science 294:1354–1357

McKay R (1997) Stem cells in the central nervous system. Science 276: 66–71

Mitsuhashi T, Aoki Y, Eksioglu YZ, Takahashi T, Bhide PG, Reeves SA, Caviness VS Jr. (2001) Overexpression of p27Kip1 lengthens the G1 phase in a mouse model that targets inducible gene expression to central nervous system progenitor cells. Proc Natl Acad Sci USA. 98:6435–6440

Sanchez-Pernaute R, Studer L, Bankiewicz KS, Major EO, McKay RD (2001) In vitro generation and transplantation of precursor-derived human dopamine neurons. J Neurosci Res 65:284–288

Mozhayeva MG, Sara Y, Liu X, Kavalali ET (2002) Development of vesicle pools during maturation of hippocampal synapses. J Neurosci 22:654–665

Nakatomi H, Kuriu T, Okabe S, Yamamoto S, Hatano O, Kawahara N, Tamura A, Kirino T, Nakafuku M (2002) Regeneration of hippocampal pyramidal neurons after ischemic brain injury by recruitment of endogenous neural progenitors. Cell 110:429–441

Nieuwkoop P (1952) Activation and organization of the amphibian central nervosu system. J Exp Zool 12: 1–130

Panchision DM, Pickel JM, Studer L, Lee SH, Turner PA, Hazel TG, McKay RD (2001) Sequential actions of BMP receptors control neural precursor cell production and fate. Genes Dev 15: 2094–2110

Petersen PH, Zou K, Hwang JK, Jan YN, Zhong W (2002) Progenitor cell maintenance requires numb and numblike during mouse neurogenesis. Nature 419:929–934

Ready D, Hanson T, Benzer S (1976) Development of the Drosophila retina, a neurocrystalline lattice. Dev Biol 53: 217–240

Renfranz PJ, Cunningham MG, McKay RD (1991) Region-specific differentiation of the hippocampal stem cell line HiB5 upon implantation into the developing mammalian brain. Cell 66: 713–729

Reubinoff BE, Pera MF, Fong CY, Trounson A, Bongso A (2000) Embryonic stem cell lines from human blastocysts: somatic differentiation in vitro. Nature Biotechnol 18:399–404

Reubinoff BE, Itsykson P, Turetsky T, Pera MF, Reinhartz E, Itzik A, Ben-Hur T (2001) Neural progenitors from human embryonic stem cells. Nature Biotechno 19:1134–1140

Reynolds BA, Weiss S (1992) Generation of neurons and astrocytes from isolated cells of the adult mammalian central nervous system. Science 255:1707–1710

Rideout WM 3rd, Hochedlinger K, Kyba M, Daley GQ, Jaenisch R (2002) Correction of a genetic defect by nuclear transplantation and combined cell and gene therapy. Cell 109:17–27

Saucedo-Cardenas O, Quintana-Hau JD, Le WD, Smidt MP, Cox JJ, De Mayo F, Burbach JP, Conneely OM (1998) Nurr1 is essential for the induction of the dopaminergic phenotype and the survival of ventral mesencephalic late dopaminergic precursor neurons. Proc Natl Acad Sci USA 95:4013–4018

Song HJ, Stevens CF, Gage FH (2002) Neural stem cells from adult hippocampus develop essential properties of functional CNS neurons. Nature Neurosci 5: 438–445

Studer L, Tabar V, McKay RD (1998) Transplantation of expanded mesencephalic precursors leads to recovery in parkinsonian rats. Nature Neurosci 1: 290–295

Studer L, Csete M, Lee SH, Kabbani N, Walikonis J, Wold B, McKay R (2000) Enhanced proliferation, survival, and dopaminergic differentiation of CNS precursors in lowered oxygen. J Neurosci 20: 7377–7783

Sun Y, Nadal-Vicens M, Misono S, Lin MZ, Zubiaga A, Hua X, Fan G,Greenberg ME (2001) Neurogenin promotes neurogenesis and inhibits glial differentiation by independent mechanisms. Cell 104:365–376

Temple S (1989) Division and differentiation of isolated CNS blast cells in microculture. Nature 340: 471–473

Thomson JA, Itskovitz-Eldor J, Shapiro SS, Waknitz MA, Swiergiel JJ, Marshall VS, Jones JM (1998) Embryonic stem cell lines derived from human blastocysts. Science 282 :1145–1147

Tornqvist N, Hermanson E, Perlmann T, Stromberg I (2002) Generation of tyrosine hydroxylase-immunoreactive neurons in ventral mesencephalic tissue of Nurr1 deficient mice. Brain Res Dev 133: 37–47

Tsai RY, McKay R (2002) Nucleolar control of cell proliferation in stem cells and cancer cells. Genes Dev 16: 2991–3003

Turnley AM, Faux CH, Rietze RL, Coonan JR, Bartlett PF (2002) Suppressor of cytokine signaling 2 regulates neuronal differentiation by inhibiting growth hormone signaling. Nature Neurosci 5:1155–1162

van Praag H, Schinder AF, Christie BR, Toni N, Palmer TD, Gage FH (2002) Functional neurogenesis in the adult hippocampus. Nature 415:1030–1034

Vicario-Abejon C, Cunningham MG, McKay RD (1995) Cerebellar precursors transplanted to the neonatal dentate gyrus express features characteristic of hippocampal neurons. J Neurosci 15: 6351–6363

Vicario-Abejon C, Collin C, Tsoulfas P, McKay RD (2000) Hippocampal stem cells differentiate into excitatory and inhibitory neurons. Eur J Neurosci 12: 677–688

Wallen AA, Zetterstrom RH, Solomin L, Arvidsson M, Olson L, Perlmann T (1999) Fate of mesencephalic AHD2-expressing dopamine progenitor cells in NURR1 mutant mice. Exp Cell Res 253:737–746

Wallen AA, Castro DS, Zetterstrom RH, karlen M, Olson L, Ericson J, Perlmann T (2001) Orphan nuclear receptor Nurr1 is essential for Ret expression in midmrain dopamine neurons and the brain stem. Mol Cell Neurosci 18:649–663

Wichterle H, Lieberam I, Porter JA, Jessell TM (2002) Directed differentiation of embryonic stem cells into motor neurons. Cell 110:385–397

Yamashita J, Itoh H, Hirashima M, Ogawa M, Nishikawa S, Yurugi T, Naito M, Nakao K, Nishikawa S (2000) Flk1-positive cells derived from embryonic stem cells serve as vascular progenitors. Nature 408:92–96

Yan J, Studer L, McKay RD (2001) Ascorbic acid increases the yield of dopaminergic neurons derived from basic fibroblast growth factor expanded mesencephalic precursors. J Neurochem 76: 307–311

Zetterstrom RH, Solomin L, Jansson L, Hoffer BJ, Olson L (1997) Dopamine neuron agenesis in Nurr1-deficient mice. Science 276 : 248–250

In Vivo Properties of In Vitro-Propagated Neural Stem Cells After Transplantation to the Neonatal and Adult Rat Brain

U. Englund[1] and A. Björklund[2]

Introduction

The ability to isolate neural stem and precursor cells and expand them in culture has provided researchers a new tool, not only assisting studies of neural development but also providing a new source of defined and expandable cells for in vivo studies using transplantation. The purposes of this chapter are, first, to review available protocols for in vitro expansion of neural precursor cells, either epigenetically using growth factors or genetically by inserting immortalizing genes; and, second, to discuss the in vivo properties of in vitro-propagated neural stem and progenitor cells, as assessed by grafting to the developing or adult rodent brain. This discussion will focus on our own recent studies using growth factor-expanded neurosphere cells of mouse and human origin and a particularly interesting, conditionally immortalized neural cell line, RN33B.

Growth factor expansion of neural progenitors

Rodent and human neural progenitors can be efficiently expanded for extended periods in culture in the presence of growth factors, both as adherent monolayers and in so-called neurosphere culture (for reviews, see Kilpatrick et al. 1995; Gage et al. 1995a; Weiss et al. 1996; Svendsen and Caldwell 2000; Martinez-Serrano et al. 2001; Ostenfeld and Svendsen 2003).

Reynolds and Weiss (1992) and Reynolds et al. (1992) were the first to successfully expand cells with stem cell-like properties from fetal and adult rodent brain, and this expansion technique has subsequently been applied to progenitors from the developing CNS of other species, including human. In their pioneering study, Reynolds and Weiss (1992) demonstrated that rodent neural progenitors derived from developing and adult forebrain can be isolated and expanded as free-floating neurospheres using EGF as mitogen. The neurosphere

[1] Present address: Dept. Neurodegenerative Disorders, H. Lundbeck A/S, Ottiliavej 9, DK-2500 Valby, Denmark

[2] Wallenberg Neuroscience Center, Dept. of Physiological Sciences, Lund University, BMC A11, S-22184 Lund, Sweden

Gage et al.
Stem Cells in the Nervous System:
Functional and Clinical Implications
© Springer-Verlag Berlin Heidelberg 2004

clusters contained undifferentiated cells, as defined by the expression of the early neural marker, nestin (Lendahl et al. 1990), and were able to differentiate into neurons, astrocytes and oligodendrocytes upon removal of the mitogen.

Soon thereafter, bFGF was shown to possess similar mitogenic effects on proliferation in both fetal and adult rodent neural multipotent progenitors, grown either as neurospheres (Temple and Qian 1995; Gritti et al. 1996; Kalyani et al. 1997; Johe et al. 1996; Qian, et al. 1997, 2000; Tropepe et al. 1999) or in adherent monolayer cultures (Ray et al. 1993; Gage et al. 1995b); Shihabuddin et al. 1997). Similar to the EGF-expanded cultures, the bFGF-stimulated cells were shown to express nestin and display multipotential differentiation upon mitogen removal. In earlier studies it appeared that bFGF was unable to maintain growth of neurospheres over multiple passages, but this was later found to be due to the absence of heparin, which is required for bFGF to act on cell aggregates (Gritti et al. 1996; Caldwell and Svendsen 1998). More recently, EGF has also been shown to be effective as a mitogen in monolayer cultures of mouse neural precursors (Eriksson et al. 2000; Skogh et al. 2001). These cultures could be expanded long-term and were shown to form both glia and neurons in the absence of EGF.

Although both EGF and bFGF are able to stimulate the proliferation of multipotent progenitors, they are likely to act on distinct precursor pools. First, there is evidence that bFGF-responsive precursors develop earlier than those responding to EGF. Thus, bFGF-responsive cells have been shown to be present from day E8.5, whereas EGF-responsive cells are detectable only from E11-E13 in the mouse forebrain (see e.g., Kilpatrick and Bartlett 1995; Johe et al. 1996; Qian et al. 1997; Kalyani et al. 1998; Tropepe et al. 1999; Represa et al. 2001). Secondly, EGF is thought to act as a mitogen for glial precursors, whereas bFGF seems to favor proliferation of neuronal precursors as well as multipotent stem/progenitor cells (Kilpatrick and Bartlett 1995; Kuhn et al. 1997; Whittemore et al. 1999). Similar to EGF, the EGF homolog TGF-α, which acts on the same receptors, has been reported to have an effect on proliferation of neural precursors similar to that of EGF. Moreover, PDGF has been shown to stimulate proliferation of rodent oligodendrocyte and type II astrocyte progenitors (Grinspan et al. 1990; Barres and Raff 1994).

Regardless of which growth factor, or growth factor combination, is used for neural stem/progenitor cell expansion, the resulting cell lines will be heterogeneous mixtures of precursors at various stages of commitment and differentiation potential. Since it is likely that the stem cell-like progenitors have a slower growth rate compared to their more committed progeny, the pool of uncommitted, multipotent cells will probably decline over time in culture. In neurospheres derived from fetal mouse or human forebrain, it has been estimated that the cells with multipotent stem cell-like properties (which are themselves able to form new neurospheres) constitute only a few percent of the total number of cells present in the spheres (see e.g., Weiss et al. 1996; Uchida et al. 2000).

Human neural progenitors have been isolated and expanded in the presence of EGF and bFGF, both as neurospheres (Brüstle et al. 1998; Flax et al. 1998; Svendsen et al. 1996, 1997, 1998; Carpenter et al. 1999; Vescovi et al. 1999a,b) and as attached cultures (Buc-Caron et al. 1995 Skogh et al. 2001). Long-term, expandable neurosphere cultures have been successfully established from 6- to 10-week-old human fetal CNS tissue. Such neurosphere cultures express nestin and, after removal of the growth factors, differentiate into neurons and glia. Svendsen and colleagues (1996, 1997, 1998) reported the generation of EGF- and/or bFGF-expanded human neurosphere cultures from the striatum, cortex and mesencephalon that could be maintained and expanded up to 150 days in vitro. Others have also successfully established multipotent human cells lines, propagated with EGF and bFGF for up to two years in culture (Flax et al. 1998; Brüstle et al. 1998; Vescovi et al. 1999a,b). A reduction in the proliferation rate after 50 days in culture has been reported for EGF- and bFGF-propagated cultures (Svendsen et al. 1996, 1997; Carpenter et al. 1999). Carpenter et al. (1999) were the first to achieve long-term expansion of EGF- and bFGF-stimulated human progenitors by the addition of leukemia inhibitory factor (LIF) to the growth expansion medium. Using a combination of these three factors enabled long-term continuous growth of multipotent progenitors for more than a year. LIF has been reported by other investigators to support the survival of neurons (Murphy and Bartlett 1993; Richards et al. 1996) and enhance the generation of neurons from human stem cells (Galli et al. 2000).

Neural stem and precursor cells are known to have high telomerase activity compared to more committed progeny and post-mitotic cells (Cai et al. 2002; Limke and Rao 2002) Interestingly, however, Ostenfeld and colleagues (2000) reported that human (but not rodent) neural progenitors have a low telomerase activity in early passage cultures. They also noted that growth factor-expanded rodent neural neurospheres that could be expanded in culture only for a limited number of passages expressed very high telomerase activity at both early and late passages, thus suggesting that even in the presence of telomerase activity alternative regulatory mechanisms are responsible for limiting the ability of rodent progenitors to undergo extended cell divisions. As noted by Ostenfeld and colleagues (2000), the lack of telomerase in late-passage human neurosphere cultures may suggest that the original stem/progenitor cell population is diluted over time by more rapidly growing and more committed progenitors with a lower telomerase activity.

In all these studies, the neurosphere cultures were established from a mixture of neural precursors at different stages of commitment. Recently, more specific populations of neural precursors have been enriched by labelling them with cell-specific antibodies or with reporter genes expressed under the control of promoters expressed in early neural precursors. Uchida et al. (2000) isolated human neural progenitor cells from fetal brain tissue by FACS using the cell surface marker CD133 and expanded the sorted cells in neurosphere culture in the presence of EGF, bFGF and LIF. The sorted cell population was greatly enriched in sphere-forming cells that, upon differentiation, could form both neurons and glia. Interestingly, most of the isolated CD133+ cells became CD133- after expansion in culture, suggesting that the cells had generated more differentiated progeny. This finding may be explained by a slower division rate of the CD133+ stem-like cells, generating more rapidly growing neural progenitors that, over time, constitute a progressively larger fraction of the neurospheres. In an alternative approach, Roy et al. (2000) and Keyoung et al. (2001) used genetic labelling to select and isolate uncommitted neural progenitors from adult and fetal human brain tissue, using adenoviral vectors expressing the GFP reporter gene under the control of promoters for two early neural genes, i.e., nestin and musashi-1 The GFP-labelled cells were sorted by FACS, and expanded as neurospheres in the presence of bFGF. The expanded cells demonstrated stem cell-like properties, i.e., self-renewal and the ability to differentiate into both neurons and glia.

Immortalized neural progenitor cell lines

Immortalized cell lines are established by transduction of precursors prior to their final mitosis with a vector encoding an immortalizing oncogene. Many laboratories have generated such cell lines, derived from various regions of the fetal or early postnatal rodent and human brain (for review, see Whittemore and Onifer 2000; Martinez-Serrano et al. 2001). There are two main types of immortalizing genes: those that encode constitutively transforming proteins [e.g., SV40 large-T antigen (large-T ag) or myc], and the temperature-sensitive (ts) mutant version of the large-T ag, which is only expressed at permissive temperatures. The two former genes, which are constitutively expressed, may retard or alter the transduced cells upon differentiation in vitro but appear to be down-regulated after transplantation (Snyder et al. 1992; Flax et al. 1998; Villa et al. 2000; Rubio et al. 2000). The advantage of using the ts large-T ag is that it will drive cell proliferation only at a reduced temperature (+33°C) but, when shifted to a non-permissive temperature (over +37°C), the genes undergo a conformational change that no longer promotes cell division. In principle, ts large-T immortalized cells should differentiate along the lineage they were committed to at the time of the infection. This may or may not be the case in cell culture: the rodent cell lines HiB5 (Renfranz et al. 1991) and ST14A (Lundberg et al. 1997) do not differentiate into mature neuronal or glial phenotypes in vitro, whereas the RN33B cell line has been found to generate neurons in vitro. Other immortalized cell lines, such as the rodent-derived MHP36 (Sinden et al. 1997), GC-B6 (Gao and Hatten 1994) and C17-2 (Snyder et al. 1997), and the human-derived H6 and HNSC.100 cell lines (Flax et al. 1998; Villa et al. 2000) have been shown to possess the capacity to differentiate into both neurons and glia in vitro. The inability of certain immortalized cell types to generate differentiated progeny in vitro could be explained by a change in potential of the originally isolated cells, caused by expansion in culture, and/or by the absence of differentiation signals normally required for maturation and survival of the immature precursors.

In vivo properties of growth factor-expanded stem/progenitor cells

Transplantation of in vitro expanded precursors to the brain of fetal or neonatal rats or mice has become a standard tool to explore their differentiation potential under conditions that are more physiological than cell culture, at least with respect to the availability of instructive signals and cell-cell interactions that are necessary to direct differentiation to more mature cellular phenotypes. In addition, several investigators have examined the capacity of such cells to integrate, migrate and differentiate after grafting to both non-neurogenic and neurogenic regions in the adult brain.

Differentiation of growth factor-expanded mouse progenitors

Previous studies have shown that bFGF-stimulated progenitors derived from adult rat hippocampus or spinal cord can differentiate into neurons in a region-specific manner after transplantation into neurogenic regions of adult recipients (the SVZ and the dentate gyrus; Gage et al. 1995b); Suhonen et al. 1996; Shihabuddin et al. 2000). Similarly, bFGF-expanded adult hippocampal progenitors, implanted into the developing rat retina, are able to generate many of the neuronal and glial cell types found in the retina (Takahashi et al. 1998).

In a first study from our laboratory, Winkler et al. (1998) transplanted mouse neural progenitors, expanded in neurosphere culture in the presence of EGF, into the forebrain ventricle in E15 rat fetuses. The transplanted cells were identified with mouse-specific antibodies, M2 and M6. In some experiments the neurospheres were generated from transgenic mice carrying *LacZ* under the nestin, GFAP- or myelin basic protein promoters. Grafted cells, as visualized with the M2/M6 antibodies, demonstrated widespread integration into telencephalic-, diencephalic- and mesencephalic structures. None of the grafted cells stained positively for nestin or vimentin or expressed the nestin-*LacZ* marker, suggesting that they did not remain in an immature state. The vast majority of the cells expressed mature glial phenotypes, most of them with the morphology of astrocytes.

In the Winkler et al. (1998) study, no signs of neurogenesis were observed from the grafted, EGF-expanded progenitors. The reason for this is not entirely clear, but it seems likely that it was due to the difficulty in visualizing, or identifying, neurons with the detection methods used in that study. In a new series of experiments (Eriksson et al. 2003), we used cells derived from an actin-GFP transgenic mouse in which the GFP transgene is well expressed in neurons in the developing and adult brain, particularly in the spine-bearing neurons of the cortex, hippocampus and striatum. Cells were taken from E13.5 LGE, MGE or cortex and expanded as neurospheres in the presence of either EGF alone, or in a combination of EGF and bFGF. Passage 5–6 neurospheres were transplanted (as intact small spheres, < 35 cells/sphere) into the hippocampus, cortex and striatum in neonatal rats. A separate group of rats received grafts of freshly dissected E13.5 LGE cells from the same transgenic mouse, for comparison. The grafted cells were visualized with a combination of the GFP marker gene, mouse-specific M2 marker and in situ hybridization for mouse satellite DNA. Only a subpopulation of the grafted cells, both neurons and glia, from either freshly dissected LGE or the expanded neurospheres, expressed GFP. Those cells that did express the transgene, on the other hand, were detected in their entirety; both the cell body and processes of neurons and glial cells, including axons, dendrites and axon terminals, were visualized in exquisite detail. Comparison with adjacent M2-stained sections showed that the distribution of M2- and GFP-positive cells overlapped, but t the GFP staining displayed many fewer cells overall. Moreover, fewer GFP-positive cells were detected at four compared to two weeks survival (whereas the number of M2-positive cells were similar at both time-points), suggesting that the GFP transgene was further down-regulated between two and four weeks post-grafting.

The transplanted mouse neurosphere cells displayed widespread migration and incorporation into all three forebrain regions. In all three sites they differentiated primarily into cells with astrocyte-like morphology, but a significant, albeit small, proportion of the cells differentiated into cells with mature neuronal morphologies. Neurons formed from the LGE-derived neurospheres were of three types: 1) cells with the morphology of medium-sized, densely spiny projection neurons in the striatum, 2) cells with interneuron-like morphologies in striatum, cortex and hippocampus, and 3) cells that integrated into the subventricular zone (SVZ) and migrated along the rostral migratory stream into the olfactory bulb. Neurosphere cells from MGE or cortex differentiated into interneuron-like cells in striatum and cortex but were not seen to either form projection neurons or integrate into the SVZ. Although the phenotypes of the differentiated neurons were not characterized in further detail in this study, the results provide evidence that growth factor-expanded mouse forebrain progenitors can generate both neurons and glia after transplantation, and that they can differentiate into mature neurons with morphological features characteristic for each target site. The results indicate, moreover, that the expanded LGE neurospheres may contain all three major types of neuronal progenitors known to be present in the E13.5 LGE, i.e. the ones giving rise to striatal projection neurons, olfactory bulb interneurons, and hippocampal/cortical interneurons, respectively, and that these progenitors may retain their regional identity over extended passages in neurosphere culture.

Extensive migration and multipotential differentiation of human neurosphere cells

Growth factor-expanded progenitors, derived from the developing forebrain of 6- to 22-week-old human fetuses, have been shown to possess a similar ability for survival, migration and differentiation as mouse neurosphere cells after implantation into the developing and adult rodent or non-human primate brain (for recent reviews, see Martinez-Serrano et al. 2001, Svendsen and Caldwell 2000; Ostenfeld and Svendsen 2003; Limke and Rao 2002).

After grafting into the forebrain ventricle in fetal or neonatal rodents, human neurosphere progenitors have demonstrated the capacity to integrate and differentiate into neurons, astrocytes and oligodendrocytes (Brüstle et al. 1998; Flax et al. 1998; Uchida et al. 2000), and human progenitors have also been reported to participate in endogenous corticogenesis after implantation into the ventricles of the developing primate forebrain (Ourednik et al. 2001). Nestin/musashi1-expressing progenitors, sorted by FACS and injected intraventricularly in utero, have been reported to integrate and generate neurons in the developing cortex (Keyoung et al. 2001). When injected into the forebrain ventricle in newborn recipients, by contrast, the same nestin/musashi1-sorted cells displayed more restricted migration and differentiated almost entirely into glial phenotypes. In addition, differentiation into neurons and glia, but more restricted migration, was reported by Rosser et al. (2000) using short-term EGF- and bFGF expanded human progenitors grafted into the striatum and hippocampus in neonatal rats. Previous transplantation experiments in adult rodents have demonstrated that in vitro-propagated human neural progenitor cells have the potential to differentiate into cells with neuron-like features. However, neurogenesis has been quite limited and variable, and it remains unclear to what extent these cells become fully differentiated and integrated with the recipient brain (see e.g., Svendsen et al. 1997; Armstrong et al. 2000, Ostenfeld et al. 2000). A critical issue is the kind of markers or detection methods used to visualize the grafted cells in their new environment. In our experience it is necessary to use a combination of species-specific, nuclear, genetic and phenotypic markers to reveal the full extent of in vivo cell differentiation from grafted human progenitors (for a detailed discussion, see Harvey 2000).

We have studied the survival, integration and differentiation of long-term EGF-, bFGF- and LIF-expanded human neural progenitor cells, grown in neurosphere culture, after transplantation to the adult brain (Fricker et al. 1999; Englund et al. 2002b) and neonatal rats (Englund et al. 2002a), using a combination of human cell-specific antisera and probes, pre-labelling with a lentiviral vector encoding GFP, and the nuclear marker BrdU for detection of the grafted cells. Both neurons and glia were formed after transplantation into striatum and hippocampus, and site-specific migration and neuronal differentiation were observed in the two classic neurogenic regions, SVZ and dentate gyrus, both in neonatal and adult recipients. The grafted human cells, as detected by an antibody specific for human nuclei, survived successfully up to 65 weeks after grafting in to the neonatal brain, without any signs of tumor formation (Englund et al. 2002a,b). Stereological assessment of the number of h-nuc-positive cells suggested that the cells had proliferated after implantation. In the neonatal hippocampal formation the total numbers of cells were about 20,000, 12,000 and 100,000, at 4, 20 and 40 weeks post-grafting, respectively, following injection of 100,000 neurosphere cells. In striatum, between 260,000 and 570,000 h nuc-positive cells were found at four weeks (200,000 cells grafted into neonates), suggesting that the human cells had divided at least once after grafting. At four weeks, staining with the human-specific proliferation marker Ki67 revealed a small number of Ki67-positive cells, predominantly located in white matter tracts (Englund et al. 2002a). This finding is consistent with previous studies suggesting that human and rodent neural progenitor cells may undergo two to three cell divisions during the first weeks after implantation (Lundberg et al. 1996; Fricker et al. 1999; Ostenfeld et al. 2000; Rosser et al. 2000; Rubio et al. 2000; Uchida et al. 2000).

Extensive migration of the grafted human progenitors, detected by the h nuclei antibody, was observed predominantly within the white matter tracts of the corpus callosum and the internal capsule, caudally along the cerebral peduncle, and rostrally along the forceps minor into the frontal cortex, after implantation into neonatal and adult recipients (Englund et al. 2002a,b). Nestin was expressed in the majority of these h nuc-positive cells, often presenting a short leading process extending parallel to the host fiber tracts. A small number of h nuc/GFAP-positive cells was also present within the corpus callosum, resembling migrating cells with processes oriented in parallel with the host white matter tracts. Some of the cells detected with the h nuc-antibody could be identified as glia-like using double immunohistochemical staining with GFAP. However, most of them had nonspecified phenotypes. None of the h nuc-positive cells located in the host white matter tracts stained positively for doublecortin (DCX), a marker for migrating and differentiating neuroblasts, suggesting that the migratory cells may be destined to form glia (GFAP/nestin-positive and DCX-negative).

Neuronal differentiation, albeit very limited, was detected after transplantation to the striatum, in both adult and neonatal recipients, as detected with antibodies to human Tau and Thy-1. hTau- and hThy-1-positive cell bodies and hTau-positive processes were found to extend rostrally and caudally within the fiber bundles of the internal capsule (Fig. 1). Some fibers from the grafted striatal neurons could be traced along the internal capsule into two of the normal striatal targets, the globus pallidus and ventral mesencephalon, suggesting that the in vitro-expanded human neural progenitors have the capacity to mature into projection neurons in the striatum. This finding is supported by the expression of some striatum-specific neuronal markers in a small fraction of the grafted cells (Fricker et al. 1999).

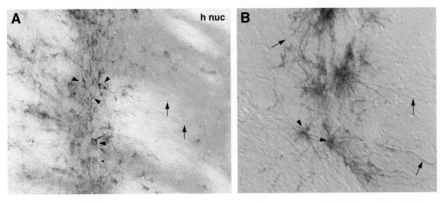

Fig. 1. Human growth factor-expanded neural progenitors transplanted to the striatum differentiated into neurons and glia, but this occurred almost exclusively at or close to the implantation site. A) hTau-positive cell bodies (arrowheads) and processes (arrows) located at the implantation site in the striatum; B) GFP-positive cells with astroglial morphologies with long processes (arrows). (From Englund et al. 2002a).

Although a significant number of the grafted human progenitors apparently responded to local cues for migration, and in part also differentiation, in the host environment, a substantial fraction of the grafted cells stayed at the implantation site. These cells expressed nestin and/ or GFAP, suggesting that they had remained undifferentiated and unable to respond to the local cues present at the graft site.

Grafting into SVZ and dentate gyrus revealed the capacity of the growth factor-expanded human progenitors to respond to the migratory cues present in these regions, both in the neonatal and adult brain. In the adult SVZ, human h nuc/DCX- double-labelled cells were found intermingled with the endogenous DCX-positive cells in a typical chain-like migratory pattern, and significant numbers of h nuc-, h nuc/DCX- as well as hTau-positive cells with profiles of migrating neurons and neuroblasts were found migrating along the RMS, towards the olfactory bulb, some of which differentiated into mature neuronal phenotypes (Fig. 2). In the dentate gyrus, h nuc- and hTau-positive cells were found predominantly in the subgranular layer and within the granular cell layer itself (Fig. 3). h nuc/DCX double-labelled neuroblasts and hTau-positive cells displayed morphologies resembling either endogenous migrating neuronal progenitors or neurons with granule cell–like morphologies with branching dendrites extending into the overlying molecular layer. Large numbers of grafted h nuc-positive cells had migrated into the molecular layer of the dentate gyrus, the pyramidal cell layer, and in the stratum radiatum and stratum oriens of the CA1-3 regions. In these regions, hTau-positive cells exhibited morphologies of mature interneurons.

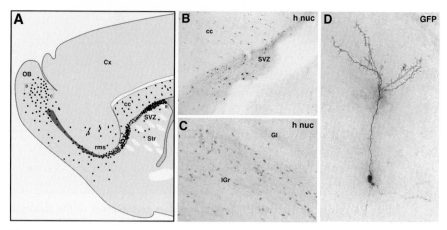

Fig. 2. Human expanded neural progenitors, grafted to the SVZ in adult rats, were able to migrate along the rostral migratory stream (A, B) into the olfactory bulb (C, D, detected with an antibody to human nuclei). In the bulb, some of the grafted cells, detected by GFP expression, were seen to differentiate into granule-like neurons, with branching dendritic trees (D). *Cx*, cortex; *cc*, corpus callosum; *gr*, granular cell layer; IGr, internal cell layer; rms, rostral migratory stream; str, striatum; SVZ, subventricular zone. (From Englund et al. 2002b).

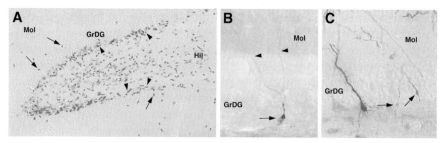

Fig. 3. Grafted human neurosphere cells, visualized with the human-specific nuclear (h nuc) antibody, migrated in large numbers into the hilus and the subgranular layer of the dentate gyrus (A, arrowheads), and a significant number of the h nuc-positive cells also appeared within the granular cell layer (A, arrows). In the granular cell layer, hTau-positive cells were observed to differentiate into cell profiles typical of granule cells, with apical dendrites oriented toward the molecular layer (B, arrows). Similarly, GFP-positive cells were found in the granule cell layer, some with morphologies of granule neurons (C, arrows). Gr, granule cell layer; Hil, hilus; Mol, molecular layer. (From Englund et al. 2002a)

Neurogenesis from immortalized neural progenitors in neonatal and adult brain

The in vivo properties of immortalized neural progenitor cell lines, as studied after transplantation to the brain, vary notably from one cell line to another (see Whittemore and Onifer 2000; Martinez-Serrano et al. 2001, for reviews). Formation of both neurons and glia has been demonstrated for a number of rodent cell lines, such as the hippocampus-derived cell lines HiB5 (Renfranz et al. 1991) and MHP36 (Sinden et al. 1997), the striatum-derived ST14A (Lundberg et al. 1997), the cerebellum-derived GC-B6 (Gao and Hatten 1994) and C17-2 (Snyder et al. 1997), and the brain stem-derived RN33B cell line (Whittemore and White 1992; Shihabuddin et al. 1995,1996; Lundberg et al. 1997). Several of these cell lines do not differentiate into neural phenotypes in vitro but exhibit multipotential maturation into both neurons and glia in vivo after transplantation. Local environmental cues present above all in the developing brain have been found to be important for neuronal differentiation of these immortalized cell lines. The hippocampus-derived HiB5 cells, for example, display site-specific neuronal and glial differentiation after implantation into the hippocampus (homotopic) and cerebellum (heterotopic) of the neonatal brain (Renfranz et al. 1991), and similar results have been obtained with the C17-2 and GC-B6 cell lines (Snyder et al. 1992; Gao and Hatten 1994). After transplantation to the adult brain, by contrast, these cells do not differentiate in a similar manner, indicating that the required signals may be down-regulated as the brain matures after birth. An exception to this rule is the RN33B cell line, which has been shown to possess a remarkable capacity to differentiate site-specifically into neurons after grafting not only intodeveloping hosts but also in the adult brain. Region-specific neuronal differentiation of RN33B cells has thus been reported in the cerebral cortex, hippocampus, spinal cord, cerebellum and striatum (Onifer et al. 1993; Shihabuddin et al. 1995, 1996; Lundberg et al. 1996). In our own recent studies (Lundberg et al. 2002; Englund et al. 2002c), we examined in further detail the ability of the RN33B cells to undergo site-specific neuronal differentiation after transplantation to different brain regions in neonatal recipients. For these experiments we used cells transduced with a lentiviral-GFP vector, which allows visualization of the morphology of the grafted cells in great detail.

The RN33B cells survived well after transplantation into all four graft sites, i.e, striatum, cortex, hippocampus and mesencephalon (Lundberg et al. 2002). At three weeks after grafting, the stereological cell counts revealed on average about 150,000 GFP-positive cells in the striatum/cortex, 70,000 in the hippocampus, and 110,000 in the mesencephalon (100,000 cells were injected in each case). Proliferation of the transplanted RN33B cells occurred after transplantation, as shown by the detection of around 150,000 GFP-positive cells in the striatum and cortex at three weeks after implantation of 100,000 cells, in agreement with a previous report using ^{3}H-thymidine, and the reporter gene *LacZ* to analyze

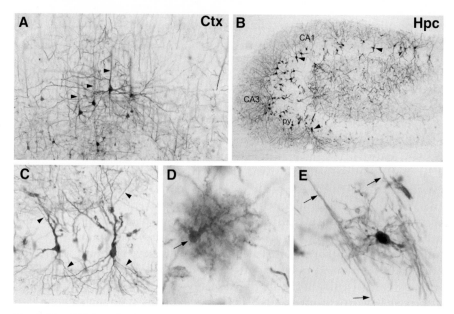

Fig. 4. The RN33B cells demonstrate a remarkable ability to differentiate into pyramidal projections neurons after grafting into neonatal cortex (Ctx, A) and hippocampus (Hpc, B, C). In the cortex, the GFP-positive grafted cells displayed morphologies typical of cortical pyramidal cells, with large pyramidal somas and spine-bearing apical and basal dendrites (A, arrowheads). Mature pyramidal neurons were also generated by GFP-expressing cells in the hippocampal CA1-3 pyramidal cell layers (B, C, arrowheads). Cells with morphological features of astrocytes (D) and oligodendrocytes (E) were also formed. (From Lundberg et al. 2002).

the grafted cells (Lundberg et al. 1996). At three weeks post-grafting, the vast majority of the GFP-expressing cells were either immature or undifferentiated. Some of these immature cells still expressed the immortalizing large T-ag gene. Many of these immature cells were associated with blood vessels or located in the ependymal or subependymal regions, and they expressed markers of neural precursors, such as nestin, vimentin and NG2. These immature cells had disappeared almost completely at longer survival times (15–17 weeks). Although this disappearance could be due to transgene down-regulation, it more likely reflects a loss of the immature cell population over time, as suggested from previous *LacZ*- and ³H-thymidine-based investigations (Lundberg et al,.1996; Shihabuddin et al. 1997).

A large fraction of the GFP-expressing cells displayed neuronal morphologies, in cortex and hippocampus (about 82% and 77% of all GFP-positive cells), at 15–17 weeks after transplantation. In contrast, a much smaller fraction developed into neurons in the striatum (~23%) and mesencephalon (~8%). Neuronal differentiation was clearly region-specific, with formation of neurons with the morphology of densely spiny projection neurons in the striatum and pyramidal neurons in the hippocampus and cortex (Fig. 4). Cells with glial morphologies, astrocytes and oligodendrocytes (Fig. 4D,E) were generally fewer, and varied markedly in number from animal to animal.

RN33B cells differentiate into fully functional pyramidal neurons in cortex and hippocampus

A large proportion of the differentiated RN33B cells in the cortex, situated in layers IV–VI, closely resembled cortical pyramidal neurons and possessed large somas and rich trees of branching and spine-bearing apical and basal dendrites (Fig. 4A,5; Lundberg et al. 2002; Englund et al. 2002c). Also in the hippocampus, the majority of GFP-expressing cells present in the pyramidal layers of CA1-CA3 displayed profiles of pyramidal neurons and possessed multiple branching dendrites, with the normal orientation perpendicular to the cell layer (Fig. 4B,C). Occasional RN33B neurons had become integrated into the granule layer of the dentate gyrus. These cells presented the characteristic morphological features of dentate granule cells, with branched apical dendrites extending into the overlying molecular layer.

Extensive GFP-positive axonal projections could be traced from the grafted neurons, as observed 15–17 weeks after transplantation. From cells located in cortex, GFP-positive projections could be traced along the corpus callosum into the contralateral cortex, and at more caudal levels within the internal capsule into the thalamus, where they formed dense fiber networks, with arborizations and terminal-like swellings, primarily in the ventrolateral and ventrobasal nuclei. The projections continued further caudally, along the cerebral peduncle into the mesencephalon, where some of these fibers were found to terminate within the substantia nigra, and occasionally down to the pons. From the grafted cells situated in the hippocampus, extensive GFP-positive projections were found throughout the ipsilateral and contralateral rostral hippocampus, with high fiber densities predominantly in the stratum oriens but also in the stratum radiatum and stratum lucidum of the CA1-3 region..

To trace the origin of the subcortically projecting axons, the retrograde tracer FluoroGold (FG) was injected into the ventrolateral and ventrobasal thalamic nuclei or the hippocampal CA3 region (Englund et al. 2002c). Consistent with the morphological observations, a significant proportion of the GFP-positive pyramidal-like cells (about 20% in cortex and 35% in hippocampus) were found to contain the FG label. In the fronto-parietal cortex, the GFP-expressing cells labeled by FG from the thalamus were found in the same layers as the FG-labeled host cells, intermingled with the intrinsic FG-positive cortico-thalamic neurons. In the hippocampus, the FG/GFP double-labelled cells occurred in the CA3 pyramidal layer.

To characterize the cortical graft-derived, pyramidal-like neurons func-tionally, we used whole-cell patch-clamp recording from individual GFP-

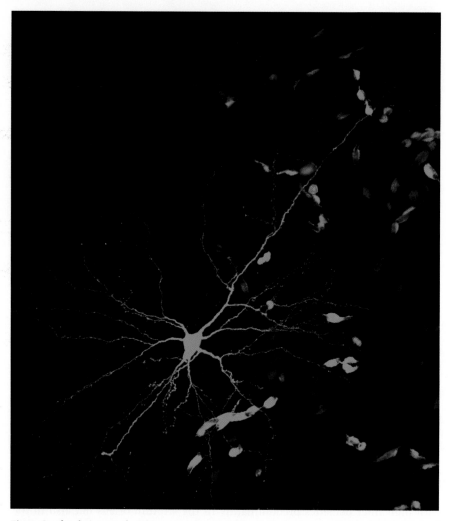

Fig. 5. Confocal picture of a GFP-expressing RN33B cell displaying a well-developed neuronal morphology, with richly spiny branching dendrites. The small, GFP-positive cells with an immature-looking morphology, mostly associated with vessels, disappeared at longer survival times.

labelled cells in cortical slices in vitro at four to seven weeks after transplantation (Englund et al. 2002c). All recorded RN33B cells were able to generate action potentials. The amplitude of action potentials, input resistance, and resting membrane potential of these cells closely resembled those of surrounding host cortical pyramidal neurons, indicating that, at this time point, the transplanted RN33B cells had reached a degree of maturation similar to that of host neurons.

Stimulation of surrounding host tissue, and simultaneously recording the electrical activity in the grafted cells, showed host excitatory and inhibitory synapses on the grafted cells. Application of receptor agonists indicated that these synapses contained functional receptors for both glutamate and GABA. These excitatory and inhibitory synapses present on the transplanted RN33B had features resembling those present on host cortical pyramidal neurons.

Together, these data demonstrate the ability of in vitro-expanded neural progenitors to develop into fully mature, functional neurons and to become synaptically integrated into host cortical circuitry, even in brain areas outside the classical neurogenic regions. At the stage of postnatal development when the RN33B cells were transplanted, cortical neurogenesis is known to be complete (Bayer and Altman 1995), whereas synaptogenesis continues well into the third postnatal week (Sutor and Luhmann 1995). Although the RN33B cells have a high intrinsic capacity to form neurons, the extent of neurogenesis and the types of neurons formed differed greatly between different brain regions.

The ability of the grafted progenitors to differentiate into diverse, and functionally mature, neuronal phenotypes and to form appropriate long-distance axonal connections is likely to depend on signals from the local tissue environment, although the factors governing their differentiation and functional integration remain largely unknown. Interestingly, in adult hosts subjected to focal excitotoxic lesions. grafted RN33B cells differentiated into fully mature neurons in areas where the intrinsic neurons were intact or partially spared, whereas cells that migrated into neuron-depleted (i.e, gliotic) areas remained undifferentiated (Shihabuddin et al. 1996; Lundberg et al. 1996). These data suggest that neuronal differentiation, and perhaps synaptogenesis and functional integration, is controlled by the presence of as yet unidentified, local cell-cell interactions rather than diffusible molecules.

Clinical perspective

More than 20 years of experimental preclinical studies in animal models of Parkinson's Disease (PD) have provided the basis for clinical trials using grafts of fetal mesencephalic dopaminergic neuroblasts in patients with PD. In the studies performed so far, tissue was taken from the developing ventral mesencephalon of human fetuses and injected stereotaxically into the putamen and/or caudate nucleus. The grafted dopaminergic neurons have been shown to survive, differentiate and establish functional contacts with the denervated host striatum, and significant and long-lasting symptomatic relief has been obtained in some, but not all, transplanted patients (see Björklund and Lindvall 2000: Brundin and Hagell 2002 for review). To date over 300 PD patients worldwide have received grafts using this technique. Although the results from the open-label clinical trials have been encouraging, it is unlikely that transplantation of primary human embryonic mesencephalic tissue can be developed into a routine therapy for many patients. This is due to the limited supply of tissue, in combination with the low survival rate of the transplanted dopamine (DA) cells (5–20%). Multiple fetal donors are needed for transplantation in each patient, a requirement that is both ethically and practically problematic (see Boer 1994; McLaren 2001; Dahlqvist 2001 for further discussion of this issue).

The problems associated with the use of primary human fetal tissue make it necessary to find alternative sources of cells for intracerebral grafting. These cells should ideally fulfill a number of requirements: 1) they should be derived from a well-characterized, reproducible and stable source; 2)they should have the potential for expansion in large numbers in vitro and allow differentiation into specific cell types upon transplantation and/or after induction in vitro; and 3) they should have proven preclinical functional efficacy in appropriate animal models. Expandable stem and progenitor cells from neural, or other, sources with the propensity to generate midbrain DA neurons are potentially interesting candidates.

Neural progenitors may, in principle, be derived from several sources: from fetal or adult brain tissue, from embryonic stem (ES) cells, or from post-mortem brain tissue. As discussed above, a number of useful protocols are now available for isolation and expansion of multipotential neural progenitors in culture, using either immortalization, epigenetic expansion and/or isolation by FACS-sorting based on cell surface markers or promoter-specific labelling. Notably, using the protocol devised by Carpenter et al. (1999), it has been possible to expand human forebrain progenitors in continuous culture up to 10^{20}-fold. A major challenge for the potential clinical use of such expandable cell lines is the identification of factors that can determine and guide differentiation of the immature cells into specific phenotypes. In principle, there are two ways of using such progenitors for transplantation in PD: the cells can either be pre-differentiated in vitro into DA neurons before grafting, or a committed precursor cell type may be identified that can differentiate into an appropriate mesencephalic dopaminergic neuronal phenotype after transplantation. All attempts using neural progenitors of human origin have so far failed to generate any significant numbers of dopaminergic neurons from this source in culture, and it has proven even more difficult to obtain any surviving, functional dopaminergic neurons in vivo after transplantation of human neurosphere cells to the striatum (see Svendsen and Caldwell 2000 for discussion).

Some so-called TH-induction protocols have been established for in vitro-expanded rodent progenitors. For example, Studer et al. (1998, 2000) published a two-step procedure for expansion of rat mesencephalic precursors in the presence of bFGF, followed by differentiation into TH-positive neurons upon removal of the mitogen. These cells were shown to survive and exhibit some functional impact after transplantation into the rat model of PD, although the survival of the expanded and predifferentiated cells was very low. Other investigators have used a mixture of cytokines, mesencephalic membrane fragments and striatal conditioned media results in stable DA differentiation of rat fetal mesencephalic precursors (Ling et al. 1998; Potter et al. 1999). Interestingly, these cells survived transplantation to the striatum in 6-OHDA-lesioned rats and provided functional benefit (Carvey et al. 2001). Human neurosphere cells, similar to those used in the present experiments, have been used to generate TH-positive neurons, at a frequency of 4–10%, by exposure to a cocktail of neurotropic factors and agents known to act on intracellular signaling pathways (Meijer et al., submitted for publication). Similarly, Storch and coworkers (2001) reported successful long-term proliferation of human mesencephalic precursors and formation of TH-positive neurons in the presence of cytokines, glial-derived growth factor and striatal co-culture. However, none or very few of these pre-differentiated TH-positive cells have been observed to survive grafting to the striatum (Meijer et al., submitted for publication).

The most promising results reported so far have been obtained with mouse ES cells. These cells can differentiate into TH-positive, DA-producing neurons either spontaneously (Björklund et al. 2002) or after pre-differentiation (Lee et al. 2000), as shown both in vitro and in vivo, and restore function after grafting into the 6-OHDA lesion model (Björklund et al. 2002; Kim et al. 2002).

None of these protocols, however, has yet been adapted for human ES cells. Nevertheless, the results obtained with progenitors derived from mouse ES cells hold promise for future development of this approach. The generally poor survival of induced human progenitors, however, suggests that it may not be trivial to translate protocols developed for mouse cells to cells of human origin. Identification of factors regulating the survival, growth and differentiation of the grafted DA neuron precursors will be important. Moreover, the role of glial cells, which may be an important constituent of primary mesencephalic grafts, in the survival, growth and differentiation of grafted dopaminergic neuron precursors needs to be explored further. Much basic work is needed to develop a better understanding of the mechanisms underlying DA neuron differentiation and survival. Without such knowledge it will be difficult to develop stem cell technology into a safe and efficient therapeutic tool for PD.

References

Armstrong RJ, Watts C, Svendsen CN, Dunnett SB, Rosser AE (2000) Survival, neuronal differentiation, and fiber outgrowth of propagated human neural precursor grafts in an animal model of Huntington's disease. Cell Transplant 9:55–64

Barres BA, Raff MC (1994) Control of oligodendrocyte number in the developing rat optic nerve. Neuron 12:935–942

Bayer SA, Altman J (1995) Neurogenesis and neuronal migration. In: Paxinos G (ed) The rat nervous system, Academic Press, San Diego, pp. 1041–1078

Björklund A, Lindvall O (2000) Cell replacement therapies for central nervous system disorders. Nature Neurosci 3:537–544

Björklund LM, Sanchez-Pernaute R, Chung S, Andersson T, Chen IY, McNaught KS, Brownell AL, Jenkins BG, Wahlestedt C, Kim KS, Isacson O (2002) Embryonic stem cells develop into functional dopaminergic neurons after transplantation in a Parkinson rat model. Proc Natl Acad Sci USA 99:1755–1757

Boer GJ (1994) Ethical guidelines for the use of human embryonic or fetal tissue for experimental and clinical neurotransplantation and research. Network of European CNS Transplantation and Restoration (NECTAR). J Neurol 242:1–13

Brundin P, Hagell P (2002) The neurobiology of cell transplantation in Parkinson's disease. Clin Neurosci Res 1:507–520

Brüstle O, Choudhary K, Karram K, Huttner A, Murray K, Dubois-Dalcq M, McKay RD (1998) Chimeric brains generated by intraventricular transplantation of fetal human brain cells into embryonic rats. Nature Biotechnol 16:1040–1044

Buc-Caron MH (1995) Neuroepithelial progenitor cells explanted from human fetal brain proliferate and differentiate in vitro. Neurobiol Dis 2:37–47

Cai J, Wu Y, Mirua T, Pierce JL, Lucero MT, Albertine KH, Spangrude GJ, Rao MS (2002) Properties of a fetal multipotent neural stem cell (NEP cell). Dev Biol 251:221–240

Caldwell MA, Svendsen CN (1998) Heparin, but not other proteoglycans potentiates the mitogenic effects of FGF-2 on mesencephalic precursor cells. Exp Neurol 152:1–10

Carpenter MK, Cui X, Hu ZY, Jackson J, Sherman S, Seiger A, Wahlberg LU (1999) In vitro expansion of a multipotent population of human neural progenitor cells. Exp Neurol 158:265–278

Carvey PM, Ling ZD, Sortwell CE, Pitzer MR, McGuire SO, Storch A, Collier TJ (2001) A clonal line of mesencephalic progenitor cells converted to dopamine neurons by hematopoietic cytokines: a source of cells for transplantation in Parkinson's disease. Exp Neurol 171:98–108

Dahlquist G (2001) Ethical guidelines for stem cell research in Sweden. Lakartidningen 19: 5890–5895

Englund U, Ericson C, Rosenblad C, Trono D, Wictorin K, Lundberg C (2000) The use of a recombinant lentiviral vector for ex vivo gene transfer into the rat CNS. Neuroreport 11: 3973–3977

Englund U, Fricker RA, Lundberg C, Björklund A, Wictorin K (2002a) Transplantation of human neural progenitor cells into the neonatal rat brain: extensive migration and differentiation with long-distance axonal projections. Exp Neurol 173:1–21

Englund U, Björklund A, Wictorin K (2002b) Migration patterns and phenotypic differentiation of long-term expanded human neural progenitor cells after transplantation into the adult rat brain. Brain Res Dev Brain Res 134:123–141

Englund U, Björklund A, Wictorin K et al, Lindvall O, Kokaia M (2002c) Grafted neural stem cells develop into functional pyramidal neurons and integrate into host cortical circuitry. Proc Natl Acad Sci USA 99: 17089–17094

Eriksson C, Ericson C, Gates MA, Wictorin K (2000) Long-term, EGF-stimulated cultures of attached GFAP-positive cells derived from the embryonic mouse lateral ganglionic eminence: in vitro and transplantation studies. Exp Neurol 164:184–199

Eriksson C, Björklund A, Wictorin K (2003) Neuronal differentiation following transplantation of expanded mouse neurosphere cultures derived from different embryonic forebrain regions. Exp Neurol, in press.

Flax JD, Aurora S, Yang C, Simonin C, Wills AM, Billinghurst LL, Jendoubi M, Sidman RL, Wolfe JH, Kim SU, Snyder EY (1998) Engraftable human neural stem cells respond to developmental cues, replace neurons, and express foreign genes. Nature Biotechnol 16:1033–1039

Fricker RA, Carpenter MK, Winkler C, Greco C, Gates MA, Bjorklund A (1999) Site-specific migration and neuronal differentiation of human neural progenitor cells after transplantation in the adult rat brain. J Neurosci 19:5990–6005

Gage FH, Ray J, Fisher LJ (1995a) Isolation, characterization, and use of stem cells from the CNS. Annu Rev Neurosci 18:159–192

Gage FH, Coates PW, Palmer TD, Kuhn HG, Fisher LJ, Suhonen JO, Peterson DA, Suhr ST, Ray J (1995b) Survival and differentiation of adult neuronal progenitor cells transplanted to the adult brain. Proc Natl Acad Sci USA 92:11879–11883

Galli R, Pagano SF, Gritti A, Vescovi AL (2000) Regulation of neuronal differentiation in human CNS stem cell progeny by leukemia inhibitory factor. Dev Neurosci 22:86-95

Gao WQ, Hatten ME (1994) Immortalizing oncogenes subvert the establishment of granule cell identity in developing cerebellum. Development 120:1059-1070

Grinspan JB, Reddy UR, Stern JL, Hardy M, Williams M, Baird L, Pleasure D (1990) Oligodendroglia express PDGF beta-receptor protein and are stimulated to proliferate by PDGF. Ann NY Acad Sci 605:71-80

Gritti A, Parati EA, Cova L, Frolichsthal P, Galli R, Wanke E, Faravelli L, Morassutti DJ, Roisen F, Nickel DD, Vescovi AL (1996) Multipotential stem cells from the adult mouse brain proliferate and self-renew in response to basic fibroblast growth factor. J Neurosci 16:1091-1100

Harvey AR (2000) Labeling and identifying grafted cells. In: Dunnett SB, Boulton AA, Baker GB (eds), Neural transplantation methods. Novel cell therapies for CNS disorders. Neuromethods, Vol. 36. Humana Press, Totowa, New Jersey, pp 319-361.

Johe KK, Hazel TG, Muller T, Dugich-Djordjevic MM, McKay RD (1996) Single factors direct the differentiation of stem cells from the fetal and adult central nervous system. Genes Dev 10: 3129-3140

Kalyani A, Hobson K, Rao MS (1997) Neuroepithelial stem cells from the embryonic spinal cord: isolation, characterization, and clonal analysis. Dev Biol 186:202-223

Keyoung HM, Roy NS, Benraiss A, Louissaint A, Jr., Suzuki A, Hashimoto M, Rashbaum WK, Okano H, Goldman SA (2001) High-yield selection and extraction of two promoter-defined phenotypes of neural stem cells from the fetal human brain. Nature Biotechnol 19:843-850

Kilpatrick TJ, Bartlett PF (1995) Cloned multipotential precursors from the mouse cerebrum require FGF-2, whereas glial restricted precursors are stimulated with either FGF-2 or EGF. J Neurosci. 15(5 Pt 1):3653-3661

Kilpatrick TJ, Richards LJ, Bartlett PF (1995) The regulation of neural precursor cells within the mammalian brain. Mol Cell Neurosci 6:2-15

Kim JH, Auerbach JM, Rodriguez-Gomez JA, Velasco I, Gavin D, Lumelsky N, Lee SH, Nguyen J, Sanchez-Pernaute R, Bankiewicz K, McKay R (2002).Dopamine neurons derived from embryonic stem cells function in an animal model of Parkinson's disease. Nature 418:50-56

Kuhn HG, Winkler J, Kempermann G, Thal LJ, Gage FH (1997) Epidermal growth factor and fibroblast growth factor-2 have different effects on neural progenitors in the adult rat brain. J Neurosci 17:5820-5829

Lee SH, Lumelsky N, Studer L, Auerbach JM, McKay RD (2000) Efficient generation of midbrain and hindbrain neurons from mouse embryonic stem cells. Nature Biotechnol 18:675-679

Lendahl U, Zimmerman LB, McKay RD (1990) CNS stem cells express a new class of intermediate filament protein. Cell 60:585-595

Limke TL, Rao MS. (2002) Neural stem cells in aging and disease. J Cell Mol Med 6(4):475-496

Ling, ZD, Potter, E, Lipton, JW, Carvey, PM (1998).Differentiation of mesencephalic progenitor cells into dopaminergic neurons by cytokines. Exp Neurol 149: 411-423

Lundberg C, Field PM, Ajayi YO, Raisman G, Björklund A (1996) Conditionally immortalized neural progenitor cell lines integrate and differentiate after grafting to the adult rat striatum. A combined autoradiographic and electron microscopic study. Brain Res 737:295-300

Lundberg C, Martinez-Serrano A, Cattaneo E, McKay RD, Björklund A (1997) Survival, integration, and differentiation of neural stem cell lines after transplantation to the adult rat striatum. Exp Neurol 145:342-360

Lundberg C, Englund U, Trono D, Björklund A, Wictorin K (2002) Differentiation of the RN33B cell line into forebrain projection neurons after transplantation into the neonatal rat brain. Exp Neurol 175:370-387

Martinez-Serrano A, Rubio FJ, Navorro B, Bueno C, Villa A (2001) Human neural stem and progenitor cells: In vitro and in vivo properties, and potential for gene therapy and cell replacemant in the CNS, Curr. Gene Therapy 1: 279-299

McLaren A (2001) Ethical and social considerations of stem cell research. Nature 414:129-131

Murphy M, Bartlett PF (1993) Molecular regulation of neural crest development. Mol Neurobiol 7: 111–135

Onifer SM, Whittemore SR, Holets VR (1993) Variable morphological differentiation of a raphe-derived neuronal cell line following transplantation into the adult rat CNS. Exp Neurol 122: 130–142

Onifer SM, Cannon AB, Whittemore SR (1997) Altered differentiation of CNS neural progenitor cells after transplantation into the injured adult rat spinal cord. Cell Transplant 6:327–338

Ostenfeld T, Caldwell MA, Prowse KR, Linskens MH, Jauniaux E, Svendsen CN (2000) Human neural precursor cells express low levels of telomerase in vitro and show diminishing cell proliferation with extensive axonal outgrowth following transplantation. Exp Neurol 164: 215–226

Ostenfeld T, Svendsen CN (2003) Recent advances in stem cell neurobiology. Adv Tech Stand Neurosurg 28:3–89

Ourednik V, Ourednik J, Flax JD, Zawada WM, Hutt C, Yang C, Park KI, Kim SU, Sidman RL, Freed CR, Snyder EY (2001) Segregation of human neural stem cells in the developing primate forebrain. Science 293:1820–1824

Potter, ED, Ling, ZD, Carvey, PM (1999) Cytokine-induced conversion of mesencephalic-derived progenitor cells into dopamine neurons. Cell Tissue Res 296: 235–246

Qian X, Davis AA, Goderie SK, Temple S (1997) FGF2 concentration regulates the generation of neurons and glia from multipotent cortical stem cells. Neuron 18:81–93

Qian X, Shen Q, Goderie SK, He W, Capela A, Davis AA, Temple S (2000) Timing of CNS cell generation: a programmed sequence of neuron and glial cell production from isolated murine cortical stem cells. Neuron 28:69–80

Ray J, Peterson DA, Schinstine M, Gage FH (1993) Proliferation, differentiation and long-term culture of primary hippocampal neurons. Proc Natl Acad Sci USA 90:3602–3606

Renfranz PJ, Cunningham MG, McKay RD (1991) Region-specific differentiation of the hippocampal stem cell line HiB5 upon implantation into the developing mammalian brain. Cell 66:713–729

Represa A, Shimazaki T, Simmonds M, Weiss S (2001) EGF-responsive neural stem cells are a transient population in the developing mouse spinal cord. Eur J Neurosci 14:452–462

Reynolds BA, Tetzlaff W, Weiss S (1992) A multipotent EGF-responsive striatal embryonic progenitor cell produces neurons and astrocytes. J Neurosci 12:4565–4574

Reynolds BA, Weiss S (1992) Generation of neurons and astrocytes from isolated cells of the adult mammalian central nervous system. Science 255:1707–1710

Richards LJ, Kilpatrick TJ, Dutton R, Tan SS, Gearing DP, Bartlett PF, Murphy M (1996) Leukaemia inhibitory factor or related factors promote the differentiation of neuronal and astrocytic precursors within the developing murine spinal cord. Eur J Neurosci 8:291–299

Rosser AE, Tyers P, Dunnett SB (2000) The morphological development of neurons derived from EGF- and FGF-2- driven human CNS precursors depends on their site of integration in the neonatal rat brain. Eur J Neurosci 12:2405–2413

Roy NS, Benraiss A, Wang S, Fraser RA, Goodman R, Couldwell WT, Nedergaard M, Kawaguchi A, Okano H, Goldman SA (2000) Promoter-targeted selection and isolation of neural progenitor cells from the adult human ventricular zone. J Neurosci Res 59:321–331

Rubio FJ, Bueno C, Villa A, Navarro B, Martinez-Serrano A (2000) Genetically perpetuated human neural stem cells engraft and differentiate into the adult mammalian brain. Mol Cell Neurosci 16:1–13

Shihabuddin LS, Hertz JA, Holets VR, Whittemore SR (1995) The adult CNS retains the potential to direct region-specific differentiation of a transplanted neuronal precursor cell line. J Neurosci 15:6666–6678

Shihabuddin LS, Brunschwig JP, Holets VR, Bunge MB, Whittemore SR (1996) Induction of mature neuronal properties in immortalized neuronal precursor cells following grafting into the neonatal CNS. J Neurocytol 25:101–111

Shihabuddin LS, Ray J, Gage FH (1997) FGF-2 is sufficient to isolate progenitors found in the adult mammalian spinal cord. Exp Neurol 148:577–586

Shihabuddin LS, Horner PJ, Ray J, Gage FH (2000) Adult spinal cord stem cells generate neurons after transplantation in the adult dentate gyrus. J Neurosci 20:8727–8735

Sinden JD, Rashid-Doubell F, Kershaw TR, Nelson A, Chadwick A, Jat PS, Noble MD, Hodges H, Gray JA (1997) Recovery of spatial learning by grafts of a conditionally immortalized hippocampal neuroepithelial cell line into the ischaemia-lesioned hippocampus. Neuroscience 81:599–608

Skogh C, Eriksson C, Kokaia M, Meijer XC, Wahlberg LU, Wictorin K, Campbell K (2001) Generation of regionally specified neurons in expanded glial cultures derived from the mouse and human lateral ganglionic eminence. Mol Cell Neurosci 17:811–820

Snyder EY, Deitcher DL, Walsh C, Arnold-Aldea S, Hartwieg EA, Cepko CL (1992) Multipotent neural cell lines can engraft and participate in development of mouse cerebellum. Cell 68: 33–51

Snyder EY, Park KI, Flax JD, Liu S, Rosario CM, Yandava BD, Aurora S (1997) Potential of neural "stem-like" cells for gene therapy and repair of the degenerating central nervous system. Adv Neurol 72:121–132

Storch A, Paul G, Csete M, Boehm BO, Carvey PM, Kupsch A, Schwarz J (2001) Long-term proliferation and dopaminergic differentiation of human mesencephalic neural precursor cells. Exp Neurol 170:317–325

Studer L, Tabar V, McKay RD (1998) Transplantation of expanded mesencephalic precursors leads to recovery in parkinsonian rats. Nature Neurosci 1:290–295

Studer L, Csete M, Lee SH, Kabbani N, Walikonis J, Wold B, McKay R (2000) Enhanced proliferation, survival, and dopaminergic differentiation of CNS precursors in lowered oxygen. J Neurosci 20:7377–7383

Suhonen JO, Peterson DA, Ray J, Gage FH (1996) Differentiation of adult hippocampus-derived progenitors into olfactory neurons in vivo. Nature 383:624–627

Sutor B, Luhmann HJ (1995) Development of excitatory and inhibitory postsynaptic potentials in the rat neocortex. Perspect Dev Neurobiol. 2(4):409–419

Svendsen CN, Clarke DJ, Rosser AE, Dunnett SB (1996) Survival and differentiation of rat and human epidermal growth factor- responsive precursor cells following grafting into the lesioned adult central nervous system. Exp Neurol 137:376–388

Svendsen CN, Caldwell MA, Shen J, ter Borg MG, Rosser AE, Tyers P, Karmiol S, Dunnett SB (1997) Long-term survival of human central nervous system progenitor cells transplanted into a rat model of Parkinson's disease. Exp Neurol 148:135–146

Svendsen CN, ter Borg MG, Armstrong RJ, Rosser AE, Chandran S, Ostenfeld T, Caldwell MA (1998) A new method for the rapid and long term growth of human neural precursor cells. J Neurosci Methods 85:141–152

Svendsen CN and Caldwell MA (2000) Stem cells in the developing central nervous system: implications for cell therapy through transplantation. In: Dunnett SB, Björklund A (eds) Functional neural transplantation II. Novel cell therapies for CNS disorders, Progress in brain research, Vol 127. Elsevier, Amsterdam, pp 13–34

Takahashi M, Palmer TD, Takahashi J, Gage FH (1998) Widespread integration and survival of adult-derived neural progenitor cells in the developing optic retina. Mol Cell Neurosci 12: 340–348

Temple S, Qian X (1995) bFGF, neurotrophins, and the control or cortical neurogenesis. Neuron 15: 249–252

Thomson JA, Itskovitz-Eldor J, Shapiro SS, Waknitz MA, Swiergiel JJ, Marshall VS, Jones JM (1998) Embryonic stem cell lines derived from human blastocysts. Science 282:1145–1147

Tropepe V, Sibilia M, Ciruna BG, Rossant J, Wagner EF, van der Kooy D (1999) Distinct neural stem cells proliferate in response to EGF and FGF in the developing mouse telencephalon. Dev Biol 208:166–188

Uchida N, Buck DW, He D, Reitsma MJ, Masek M, Phan TV, Tsukamoto AS, Gage FH, Weissman IL (2000) Direct isolation of human central nervous system stem cells. Proc Natl Acad Sci USA 97: 14720–14725

Vescovi AL, Gritti A, Galli R, Parati EA (1999a) Isolation and intracerebral grafting of nontransformed multipotential embryonic human CNS stem cells. J Neurotrauma 16:689–693

Vescovi AL, Parati EA, Gritti A, Poulin P, Ferrario M, Wanke E, Frolichsthal-Schoeller P, Cova L, Arcellana-Panlilio M, Colombo A, Galli R (1999b) Isolation and cloning of multipotential stem cells from the embryonic human CNS and establishment of transplantable human neural stem cell lines by epigenetic stimulation. Exp Neurol 156:71–83

Villa A, Snyder EY, Vescovi A, Martinez-Serrano A (2000) Establishment and properties of a growth factor-dependent, perpetual neural stem cell line from the human CNS. Exp Neurol 161:67–84

Weiss S, Dunne C, Hewson J, Wohl C, Wheatley M, Peterson AC, Reynolds BA (1996) Multipotent CNS stem cells are present in the adult mammalian spinal cord and ventricular neuroaxis. J Neurosci 16:7599–7609

Whittemore SR, Morassutti DJ, Walters WM, Liu RH, Magnuson DS (1999) Mitogen and substrate differentially affect the lineage restriction of adult rat subventricular zone neural precursor cell populations. Exp Cell Res 252:75–95

Whittemore SR, Onifer SM (2000) Immortalized neural cells for CNS transplantation. In: Dunnett SB,Björklund A (eds) Functional Neural transplantation II. Novel cell therapies for CNS disorders, Progress in brain research, Vol 127. Elsevier, Amsterdam, pp 49–66

Winkler C, Fricker RA, Gates MA, Olsson M, Hammang JP, Carpenter MK, Bjorklund A (1998) Incorporation and glial differentiation of mouse EGF-responsive neural progenitor cells after transplantation into the embryonic rat brain. Mol Cell Neurosci 11:99–116

Neurogenesis in Stroke and Epilepsy

Z. Kokaia[1], A. Arvidsson[1], C. Ekdahl[1] and O. Lindvall[1]

Summary

Ischemic and epileptic insults promote neurogenesis from neural stem cells located in the dentate subgranular zone and in the subventricular zone lining the lateral ventricles. These findings raise the possibility that the adult brain tries to use its own stem cells to repair itself. New neurons generated by ischemic insults have now been shown to migrate from the subventricular zone and posterior periventricle to the damaged striatum and CA1 region, respectively. Here they express morphological markers characteristic of those neurons that have died, and some evidence indicates that these neurons can re-establish connections. However, we still lack much information about the regulation of insult-induced neurogenesis and its behavioral consequences. The new neurons may contribute to functional recovery but have also been suggested to be involved in the development of epilepsy.

Introduction

In the adult brain, neurogenesis from neural stem cells and progenitor cells continues in two regions: the subventricular zone (SVZ), which lines the lateral ventricles and gives rise to new interneurons that reach the olfactory bulb via the rostral migratory stream (RMS), and the subgranular zone (SGZ) of the dentate gyrus (DG), which generates new granule cells (Gage 2000). Additional neuronal progenitors reside in the forebrain parenchyma (Reynolds and Weiss 1992). The interest in neurogenesis and brain insults started in 1997, when it was demonstrated that brief periods of seizure activity as well as status epilepticus stimulated the formation of new neurons in the rat dentate gyrus (Bengzon et al. 1997; Parent et al. 1997). It was soon found that ischemic insults also promoted neurogenesis in the dentate subgranular zone (Liu et al. 1998), and recently, Parent and co-workers (2002a) showed that status epilepticus enhances the formation of new neurons in the SVZ-RMS. Much research in this field has been

[1] Section of Restorative Neurology, Wallenberg Neuroscience Center, University Hospital BMC A-11, SE-221 84 Lund, Sweden

Gage et al.
Stem Cells in the Nervous System:
Functional and Clinical Implications
© Springer-Verlag Berlin Heidelberg 2004

focused on two main questions. First, because these insults are associated with neuronal death, can the new neurons replace those that have died due to the insult? More specifically, does neuronal self-repair exist in the adult brain after such insults? From a clinical perspective, this is of course a particularly exciting possibility. Second, how are the different steps of insult-induced neurogenesis (proliferation, migration, differentiation and survival) regulated?

In this chapter, we will describe recent findings clearly supporting the idea that the adult brain tries to use its own brain cells to repair itself after ischemic insults. We will also summarize some of the experimental evidence, mainly from ischemia studies, regarding the regulation and functional consequences of insult-induced neurogenesis.

Does neuronal self-repair occur after insults to the adult brain?

The evidence that new neurons formed from neural stem cells in the adult brain can replace neurons that have died after injury has been scarce until recently. (Magavi et al. 2000) used targeted apoptosis of cortical pyramidal neurons in mice and found a small number of new neurons extending processes to the original target sites in the thalamus. This lesion only destroyed the targeted neurons, without affecting the surrounding tissue. However, during the past few months, three papers have been published that indicate that ischemic insults, leading to much more extensive damage, give rise to neuronal self-repair (Arvidsson et al. 2002; Nakatomi et al. 2002; Parent et al. 2002b).

Stroke induced by middle cerebral artery occlusion (MCAO) leads to increased cell proliferation and increased numbers of immature neurons in the ipsilateral SVZ (Jin et al. 2001; Zhang et al. 2001; Arvidsson et al. 2002; Parent et al. 2002b; Takasawa et al. 2002). New neurons appear to migrate to the damaged striatal area (Arvidsson et al. 2002; Parent et al. 2002b). At two weeks following the insult, large numbers of migrating neuroblasts extend in a gradient from the SVZ laterally up to 2 mm into the ischemic striatum. The majority of these neurons had been formed by proliferation after the stroke, but neuroblasts that had undergone division prior to the insult were also recruited to the damaged area (Arvidsson et al. 2002). In addition, new neurons could have originated from progenitors in the striatal parenchyma.

At two weeks after the stroke, virtually all of the new cells expressed Meis2 and Pbx proteins (Arvidsson et al. 2002), which are normally co-localized within developing striatal, medium-sized spiny neurons (Toresson et al. 2000). A longer survival time allowed for the proliferated cells to differentiate into mature neurons. At five to six weeks after stroke, the new neurons were found mainly in the damaged area but also in the unaffected medial striatum. Many of the new neurons expressed DARPP-32 or calbindin (Arvidsson et al. 2002; Parent et al. 2002b), i.e., two markers of striatal, medium-sized spiny neurons.

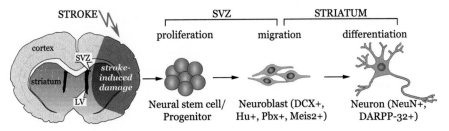

Fig. 1. Schematic representation of stroke-induced neurogenesis in the striatum. Neural stem or progenitor cells reside in the subventricular zone (SVZ). Stroke that causes pronounced loss of striatal and cortical neurons gives rise to increased cell proliferation in the SVZ. Neuroblasts formed after and to some extent also prior to the stroke then migrate to the damaged part of the striatum, where they start to express markers specific for striatal projection neurons. LV, lateral ventricle.

Thus, stroke induces the new neurons to differentiate into the phenotype of the majority of the dead neurons.

Using a different type of ischemic insult, i.e., global forebrain ischemia evoked by transient occlusion of the carotid and vertebral arteries, Nakafuku and co-workers (Nakatomi et al. 2002) found that intraventricular infusion of FGF-2 and EGF for four days after the ischemia gave rise to substantial regeneration of hippocampal CA1 pyramidal neurons in rats. The magnitude of this neuroregenerative response is still unclear because cell counting was performed in single, thin brain sections using non-stereological procedures. The new CA1 neurons were derived from progenitor cells around the posterior periventricle adjacent to the hippocampus (Nakatomi et al. 2002). These cells proliferated in response to ischemia and were traced to the CA1 region, where they seemed to receive synaptic input and project to subiculum.

Whether new neurons can be formed in the adult cerebral cortex after brain insults leading to extensive damage is controversial. (Gu et al. 2000) found a small proportion of proliferated, BrdU-positive cells co-labeled with neuronal markers in the penumbra zone in a stroke model using a phototrombotic ring. The same group subsequently reported the presence of double-labelled cells in the cerebral cortex, particularly in the peri-infarct area, after MCAO (Jiang et al. 2001). However, three other studies did not detect any BrdU-positive cells co-labelled with neuronal markers in the injured cortex following MCAO (Zhang et al. 2001; Arvidsson et al. 2002; Parent et al. 2002b). Taken together, these data indicate that the instructive or survival cues necessary for neurogenesis are lacking or that this process is suppressed in the ischemically damaged cortex.

How is neurogenesis regulated after brain insults?

The molecular mechanisms regulating insult-induced neurogenesis are only partly understood. Hypothetically, neurogenesis following brain insults could proceed as it does during embryonic development, involving the same concerted action of transcription factors, signaling molecules, and growth factors. Accordingly, both the stroke-generated striatal cells (Arvidsson et al. 2002) and the cells in the posterior periventricle that had proliferated in response to global ischemia (Nakatomi et al. 2002) initially expressed several developmental transcription factors.

Recently, Jin et al. (2002) reported that hypoxia in cortical cultures triggered neurogenesis, which was mediated through increased FGF-2 and stem cell factor (SCF) levels. These findings suggest that ischemic insults stimulate neurogenesis in the SVZ and SGZ in vivo through the release of SCF and/or FGF-2. In support of this hypothesis, the increase of neurogenesis in the SGZ following MCAO was found to be attenuated in homozygous FGF-2 knockout mice (Yoshimura et al. 2001). Jin et al. (2002) also found increased expression of the SCF receptor, c-kit, in the SVZ and SGZ after MCAO. Intraventricular administration of SCF gave rise to increased numbers of proliferated cells co-expressing a neuronal marker in the SVZ and SGZ.

Intraventricular infusion of BDNF protein (Zigova et al. 1998; Pencea et al. 2001) or overexpression of the BDNF gene in the ventricular zone (Benraiss et al. 2001) in intact, adult rats increase the number of new neurons in the RMS and olfactory bulb, striatum, septum, thalamus and hypothalamus. These findings raise the possibility that BDNF is a neurogenesis-promoting factor after ischemic brain insults as well. However, currently this is a controversial issue. Thus, (Larsson et al. 2002) found that long-term delivery of BDNF to the hippocampus counteracts neuronal differentiation but not cell proliferation and survival of newly formed cells in the DG after global forebrain ischemia.

Erythropoietin (EPO) is another factor that probably modulates neurogenesis following ischemia (Shingo et al. 2001). This cytokine is produced as part of the ischemic-hypoxic response, and EPO receptors are expressed in the SVZ. Intraventricular infusion of EPO and EPO antibodies (Shingo et al. 2001) leads to increased and decreased production, respectively, of olfactory bulb neurons from SVZ.

Glutamatergic mechanisms are involved in the regulation of SGZ neurogenesis after global (Bernabeu and Sharp 2000) and focal (Arvidsson et al. 2001) ischemia. Both NMDA and AMPA receptor blockade at the time of global ischemia prevented the increase of neurogenesis as well as the CA1 neuronal death. However, MCAO that did not lead to ischemic hippocampal damage gave rise to increased neurogenesis, which was completely suppressed by NMDA but not by AMPA receptor blockade (Arvidsson et al. 2001). These findings suggest that the mechanisms modulating SGZ neurogenesis are at least partly different following MCAO and global ischemia. We have proposed (Arvidsson et al. 2001) that the marked stimulation of neurogenesis by NMDA receptor activation following stroke is mediated by the glutamate-evoked increased levels of certain growth factors, e. g., FGF-2 and BDNF.

It will be important to clarify which signalling molecules attract the new neurons and lead to their migration to the damaged area. Such signals could also explain why transplanted neuronal precursors and neuroepithelial stem cells migrate to the tumor (Aboody et al. 2000) or ischemic lesion (Veizovic et al. 2001), respectively, in the contralateral hemisphere, and also why stem cells grafted at the border of the striatal lesion following MCAO preferentially migrate into the damaged area (unpublished observation). In the striatum, the new neurons sometimes form aggregates in close contact with astrocytes, suggesting that some of them may reach the damaged area using so-called chain migration, similar to neuronal precursors in the RMS (Arvidsson et al. 2002; Parent et al. 2002b). The demonstration of an astrocyte-derived migration-inducing activity (Mason et al. 2001), regulating the migration of SVZ neuronal precursors, suggests that astrocytes in the ischemic lesion (Stoll et al. 1998) may play a similar role. Whether guidance cues that are involved in the outward migration of SVZ cells in the striatal primordium during development (Hamasaki et al. 2001) play a role in the adult, ischemia-damaged striatum is unknown.

The vast majority of the new striatal neurons die between two and five weeks after the stroke (Arvidsson et al. 2002), a time frame that resembles the loss of SGZ cells that have proliferated after status epilepticus (Ekdahl et al. 2001). This degeneration probably reflects an unfavourable environment for the newly formed neurons, a lack of appropriate trophic support and connections, and exposure to the detrimental action of damaged tissue. The seizure-generated hippocampal neurons die, at least partly, through a caspase-mediated apoptotic pathway (Ekdahl et al. 2001, 2002). It remains to be explored whether similar death mechanisms operate in neurons generated after ischemic insults.

What are the functional consequences of insult-induced neurogenesis?

Whether the new neurons generated after brain insults contribute to functional recovery or, conversely, have adverse effects is still virtually unknown. The role of the new striatal neurons generated after stroke has not yet been studied. Although the animals subjected to global forebrain ischemia and EGF+FGF-2 infusion were reported to exhibit improved hippocampal synaptic transmission and performance in a learning and memory task (Nakatomi et al. 2002), it cannot be excluded, based on available data, that these effects were caused by the growth factors per se and not by the neurogenesis. Labelling with a GFP-retroviral vector followed by anatomical and patch-clamp analyses (van Praag et al. 2002) and virus-based transsynaptic neuronal tracing and c-fos mapping of neuronal activity (Carlén et al. 2002) could be useful tools for determining the functional properties and integration into host neural circuitries of cells generated after insults.

Stroke is often followed by spontaneous recovery. In rats, only a small fraction of the dead striatal neurons (about 0.2%) is replaced by the new neurons (Arvidsson et al. 2002). Whether the new neurons contribute to the improvement of sensorimotor function is unclear. To explore the functional consequences of striatal neurogenesis and the causative link between hippocampal neurogenesis and recovery of learning and memory following cerebral ischemia, the effects of suppression of neurogenesis in the post-ischemic period should be assessed.

One hypothesis has been that the neurons formed after seizures could contribute to the development of epilepsy. In agreement with this hypothesis, Scharfman et al. (2000) have reported that cells that have proliferated after status epilepticus are found ectopically in the dentate hilus. These granule-like cells exhibit pro-epileptogenic burst firing.

Conclusions

The studies reviewed here provide experimental evidence that the cellular plasticity in the adult brain is substantial and that neuronal self-repair by recruitment of endogenous precursors may occur after insults. However, we lack a lot of information about the functional properties of the generated cells and the behavioral consequences of neurogenesis. The mechanisms triggering increased cell proliferation and regulating survival and migration of progenitor cell progeny and its differentiation into specific neuron types are virtually unknown. Of major importance is to develop strategies to optimize the neuronal replacement with respect to both the number of surviving neurons and their anatomical and functional integration into the host brain. Neurogenesis may contribute to the development of epilepsy but the most exciting possibility is if neurogenesis can be used to repair the damaged brain. Obviously, the self-repair mechanisms described here are insufficient and lead to incomplete functional recovery. However, the data obtained raise the possibility that amplification of self-repair mechanisms might in the future be of therapeutic value for brain-injured patients.

Acknowledgments

Our own research was supported by the Swedish Research Council, The Söderberg, Kock, Crafoord, and Elsa and Thorsten Segerfalk Foundations, the Swedish Stroke Foundation, and the Swedish Association of Neurologically Disabled.

References

Aboody KS, Brown A, Rainov NG, Bower KA, Liu S, Yang W, Small JE, Herrlinger U, Ourednik V, Black PM, Breakefield XO, Snyder EY (2000) Neural stem cells display extensive tropism for pathology in adult brain: evidence from intracranial gliomas. Proc Natl Acad Sci USA 97: 12846–12851

Arvidsson A, Kokaia Z, Lindvall O (2001) N-methyl-D-aspartate receptor-mediated increase of neurogenesis in adult rat dentate gyrus following stroke. Eur J Neurosci 14:10–18

Arvidsson A, Collin T, Kirik D, Kokaia Z, Lindvall O (2002) Neuronal replacement from endogenous precursors in the adult brain after stroke. Nature Med 8:963–970

Bengzon J, Kokaia Z, Elmer E, Nanobashvili A, Kokaia M, Lindvall O (1997) Apoptosis and proliferation of dentate gyrus neurons after single and intermittent limbic seizures. Proc Natl Acad Sci USA 94:10432–10437

Benraiss A, Chmielnicki E, Lerner K, Roh D, Goldman SA (2001) Adenoviral brain-derived neurotrophic factor induces both neostriatal and olfactory neuronal recruitment from endogenous progenitor cells in the adult forebrain. J Neurosci 21:6718–6731

Bernabeu R, Sharp FR (2000) NMDA and AMPA/kainate glutamate receptors modulate dentate neurogenesis and CA3 synapsin-I in normal and ischemic hippocampus. J Cereb Blood Flow Metab 20:1669–1680

Carlen M, Cassidy RM, Brismar H, Smith GA, Enquist LW, Frisen J (2002) Functional integration of adult-born neurons. Curr Biol 12:606–608

Ekdahl CT, Mohapel P, Elmer E, Lindvall O (2001) Caspase inhibitors increase short-term survival of progenitor-cell progeny in the adult rat dentate gyrus following status epilepticus. Eur J Neurosci 14:937–945

Ekdahl CT, Mohapel P, Weber E, Bahr B, Blomgren K, Lindvall O (2002) Caspase-mediated death of newly formed neurons in the adult rat dentate gyrus following status epilepticus. Euro J Neurosci, 16:1463–1471

Gage F (2000) Mammalian neural stem cells. Science 287:1433–1438

Gu W, Brännström T, Wester P (2000) Cortical neurogenesis in adult rats after reversible photothrombotic stroke. J Cereb Blood Flow Metab 20:1166–1173

Hamasaki T, Goto S, Nishikawa S, Ushio Y (2001) A role of netrin-1 in the formation of the subcortical structure striatum: repulsive action on the migration of late-born striatal neurons. J Neurosci 21:4272–4280

Jiang W, Gu W, Brannstrom T, Rosqvist R, Wester P (2001) Cortical neurogenesis in adult rats after transient middle cerebral artery occlusion. Stroke 32:1201–1207

Jin K, Minami M, Lan JQ, Mao XO, Batteur S, Simon RP, Greenberg DA (2001) Neurogenesis in dentate subgranular zone and rostral subventricular zone after focal cerebral ischemia in the rat. Proc Natl Acad Sci USA 98:4710–4715

Jin K, Mao XO, Sun Y, Xie L, Greenberg DA (2002) Stem cell factor stimulates neurogenesis in vitro and in vivo. J Clin Invest 110:311–319

Larsson E, Mandel RJ, Klein RL, Muzyczka N, Lindvall O, Kokaia Z (2002) Suppression of insult-induced neurogenesis an adult rat brain by brain-derived neurotrophic factor. Exp Neurol 177: 1–8

Liu J, Solway K, Messing RO, Sharp FR (1998) Increased neurogenesis in the dentate gyrus after transient global ischemia in gerbils. J Neurosci 18:7768–7778

Magavi SS, Leavitt BR, Macklis JD (2000) Induction of neurogenesis in the neocortex of adult mice. Nature 405:951–955

Mason HA, Ito S, Corfas G (2001) Extracellular signals that regulate the tangential migration of olfactory bulb neuronal precursors: inducers, inhibitors, and repellents. J Neurosci 21: 7654–7663

Nakatomi H, Kuriu T, Okabe S, Yamamoto S, Hatano O, Kawahara N, Tamura A, Kirino T, Nakafuku M (2002) Regeneration of hippocampal pyramidal neurons after ischemic brain injury by recruitment of endogenous neural progenitors. Cell 110:429–441

Parent JM, Yu TW, Leibowitz RT, Geschwind DH, Sloviter RS, Lowenstein DH (1997) Dentate granule cell neurogenesis is increased by seizures and contributes to aberrant network reorganization in the adult rat hippocampus. J Neurosci 17:3727–3738

Parent JM, Valentin VV, Lowenstein DH (2002a) Prolonged seizures increase proliferating neuroblasts in the adult rat subventricular zone-olfactory bulb pathway. J Neurosci 22:3174–3188

Parent JM, Vexler ZS, Gong C, Derugin N, Ferriero DM (2002b) Rat forebrain neurogenesis and striatal neuron replacement after focal stroke. Ann Neurol 52:802–813

Pencea V, Bingaman KD, Wiegand SJ, Luskin MB (2001) Infusion of brain-derived neurotrophic factor into the lateral ventricle of the adult rat leads to new neurons in the parenchyma of the striatum, septum, thalamus, and hypothalamus. J Neurosci 21:6706–6717

Reynolds BA, Weiss S (1992) Generation of neurons and astrocytes from isolated cells of the adult mammalian central nervous system. Science 255:1707–1710

Scharfman HE, Goodman JH, Sollas AL (2000) Granule-like neurons at the hilar/CA3 border after status epilepticus and their synchrony with area CA3 pyramidal cells: functional implications of seizure-induced neurogenesis. J Neurosci 20:6144–6158

Shingo T, Sorokan ST, Shimazaki T, Weiss S (2001) Erythropoietin regulates the in vitro and in vivo production of neuronal progenitors by mammalian forebrain neural stem cells. J Neurosci 21:9733–9743

Stoll G, Jander S, Schroeter M (1998) Inflammation and glial responses in ischemic brain lesions. Prog Neurobiol 56:149–171

Takasawa K, Kitagawa K, Yagita Y, Sasaki T, Tanaka S, Matsushita K, Ohstuki T, Miyata T, Okano H, Hori M, Matsumoto M (2002) Increased proliferation of neural progenitor cells but reduced survival of newborn cells in the contralateral hippocampus after focal cerebral ischemia in rats. J Cereb Blood Flow Metab 22:299–307

Toresson H, Parmar M, Campbell K (2000) Expression of Meis and Pbx genes and their protein products in the developing telencephalon: implications for regional differentiation. Mech Dev 94:183–187

van Praag H, Schinder AF, Christie BR, Toni N, Palmer TD, Gage FH (2002) Functional neurogenesis in the adult hippocampus. Nature 415:1030–1034

Veizovic T, Beech JS, Stroemer RP, Watson WP, Hodges H (2001) Resolution of stroke deficits following contralateral grafts of conditionally immortal neuroepithelial stem cells. Stroke 32:1012–1019

Yoshimura S, Takagi Y, Harada J, Teramoto T, Thomas SS, Waeber C, Bakowska JC, Breakefield XO, Moskowitz MA (2001) FGF-2 regulation of neurogenesis in adult hippocampus after brain injury. Proc Natl Acad Sci USA 98:5874–5879

Zhang RL, Zhang ZG, Zhang L, Chopp M (2001) Proliferation and differentiation of progenitor cells in the cortex and the subventricular zone in the adult rat after focal cerebral ischemia. Neuroscience 105:33–41

Zigova T, Pencea V, Wiegand SJ, Luskin MB (1998) Intraventricular administration of BDNF increases the number of newly generated neurons in the adult olfactory bulb. Mol Cell Neurosci 11:234–245

The Incorrect Use of Transcription Factors: A Key to Your (STEM) Cells?

A. Prochiantz[1], G. Mainguy[1], L. Sonnier[1], I. Caillé[1],
B. Lesaffre[1], M. Volovitch[1] and A. Joliot[1]

Summary

The presence of stem cells in the adult, in particular but not only in the central nervous system, suggests that one could use this reservoir for replacement therapies. Thus large numbers of studies are aimed at elucidating the sequence of signaling events leading from stem cells to the cells with the desired differentiated phenotype.

Recently, we have observed that homeoprotein transcription factors have the uncanny property of transferring between cells, and the sequences and mechanisms involved in the unconventional secretion and uptake of these transcription factors have been elucidated, at least in part.

Based on this newly acquired knowledge, we have been able to address introduce exogenous molecules, including transcription factors, into live cells. We propose that this technology could be used to amplify and reprogram stem cells in vitro and in vivo.

In this short review we shall describe the present knowledge concerning transcription factor intercellular exchange and discuss the physiological and technological consequences of the phenomenon.

Introduction

Within a few years the representation of the adult nervous system has dramatically evolved to include the idea of neuronal renewal. Indeed since the pioneering studies of Nottebhom and colleagues (Alvarez-Buylla et al. 1988; Nottebohm 2002) on the making of new neurons in the song center of the zebra-finch, many groups have contributed to the demonstration that stem cells exist in most species, primates included (Gould et al. 1999; Clarke et al. 2000; Bernier et al. 2002; Kornack and Rakic 1999, 2001). Although neurogenesis seems to be most active in a few zones, for example the dentate gyrus (hippocampus) or the sub-ventricular zone (SVZ), the phenomenon is probably more general.

---

[1] Ecole Normale Supérieure and CNRS UMR8542, 46 rue d'Ulm, 75005 Paris, France,
  Mail: prochian@wotan.ens.fr

Gage et al.
Stem Cells in the Nervous System:
Functional and Clinical Implications
© Springer-Verlag Berlin Heidelberg 2004

This finding has raised the hope that replacement therapies based on neuronal or glial renewal could be developed in the near future (Clarke et al. 2000; Nottebohm 2002).

To that end several strategies can be envisaged. One is to start with embryonic stem (ES) cells in vitro and drive them into specific differentiation pathways. For example, programs have been initiated that aim at favoring the differentiation of ES cells into specific classes of neurons, such as motor neurons or dopaminergic (DA) neurons (Wichterle et al. 2002; Kim J-H et al. 2002). These cells could then be grafted into the brain of patients.

Another approach is to take the cells in the brain, for example at the level of the SVZ, expand them in vitro and differentiate them along the appropriate pathways (Lois and Alvarez-Buylla 1993).

Finally, the most interesting approach would consist in amplifying and programming stem cells in vivo by non-invasive technologies and force them to repopulate the target cerebral zones (Horner and Gage 2000; Kruger and Morrison 2002; Kuhn and Svendsen 1999).

All these approaches underscore the necessity to elucidate the mechanisms at the basis of stem cell (or progenitor) proliferation and differentiation into defined pathways. From this point of view, the studies on neural induction (ES cells derive from the epiblast) and on lineages are illuminating because they have already contributed to the identification of growth and transcription factors participating in the latter processes (Munoz-Sanjuana and Brivanlou 2002; Shilo 2001).

In fact, transcription factors are of particular interest because they transduce the instruction of the growth factors at the transcription level. Thus one possible access to the regulation of stem cell/progenitor proliferation or differentiation would be to directly introduce transcription factors or transcription factor regulators into live cells.

From this point of view, the finding that homeoprotein transcription factors are captured by live cells and addressed to their cytoplasm and nucleus might be of therapeutic value (Prochiantz 2000).

Homeoproteins Travel

The concept of messenger proteins takes its origin in the observation that several homeoproteins are both secreted and internalized by live cells (Prochiantz 2000). Although the significance of this phenomenon is still unclear, the mechanisms for unconventional secretion and internalization have been elucidated, in part. Internalization is driven by the third helix of the homeodomain, a short and highly conserved sequence of 16 amino acids (Derossi et al. 1994,1996). In fact this peptide and other peptides derived from it (the penetratin family of cell-permeable peptides) not only cross biological membranes (likely through the induction of inverted micelles) but also carry several types of covalently attached cargoes into live cells (Derossi et al. 1998; Dupont et al. 2002). The nature and size of cargoes range from small oligo-nucleotides and peptides (or phosphopeptides) to high molecular weight proteins. Following internalization, these cargoes gain access to the cell cytoplasm and nucleus, where they can interfere with intracellular physiological functions, including signal transduction and genes transcription.

Homeoproteins are not only internalized but also secreted, in spite of the absence of a signal peptide (Joliot et al. 1998). Secretion requires a small peptide sequence distinct from the internalization sequence and with nuclear export properties (Maizel et al. 1999). This sequence differs from the internalization one but is also present in the homeodomain. Secretion involves a passage into a secretory compartment independently of the presence of a signal peptide and is regulated by the phosphorylation of a serine-rich domain upstream of the homeodomain (Maizel et al. 2002).

Homeoprotein-Derived Vectors

The fact that the third helix of the Antennapedia homeodomain (and of most homeodomains) is necessary and sufficient for translocation allowed us to address the mechanism of translocation. This third helix was synthesized, and its internalization in live cells was followed, thanks to a biotin residue covalently attached to its N-terminus. Translocation occurs at 4°C and 37°C and is not concentration-dependent between 10 pM and 100 µM (Dupont et al. 2002). Toxicity is rare below 10 µM, and the presence of positively charged amino acids like arginines and lysines and hydrophobic residues confers to the peptide an isoelectric point above 12 and amphipatic characteristics. Circular dichroïsm experiments have shown that the peptide is poorly structured in water, but adopts an alpha-helical structure in a hydrophobic environment (Derossi et al. 1994).

Internalization at 4°C suggested the absence of a chiral receptor. To verify this point, we synthesized a peptide entirely composed of D-amino acids that was incapable of stereospecific binding to a receptor. This peptide was also internalized, thus demonstrating the absence of chiral receptor (Derossi et al. 1996). Starting from the sequence of the third helix of the Antennapedia homeodomain, a number of peptides defining the "penetratin family" were developed. Studying these peptides and also some variants that have lost the ability to transfer across biological membranes led to the proposal and partial demonstration that internalization is a two-step process. A first step would consist in the electrostatic association of the peptides to negatively charged microdomains of the membrane; the second step would consist in the destabilization of the membrane and the formation of an inverted micelle. In our model, the latter destabilization requires a tryptophan residue as replacing the tryptophan in position 48 of the homeodomain by a phenylalanin blocks

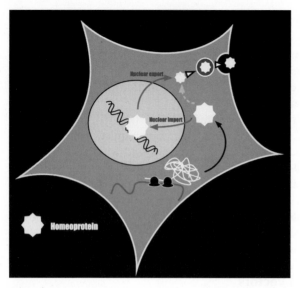

Fig. 1. Homeoprotein secretion.In the model developed in the laboratory for Engrailed, the homeoprotein synthesized in the cytoplasm is transported to the nucleus thanks to its nuclear localization signal. Shuttling between the nucleus and cytoplasm is permitted by the presence of a nuclear export sequence that is also necessary for secretion. From this observation and also from several experimental results correlating nuclear export and secretion, it can be proposed that secretion necessitates a passage through the nucleus. It is also important to notice that secretion requires access of the homeoprotein to a class of vesicles that have been characterized as enriched in cholesterol and glycosphingolipids. This is in contrast with internalization, which does not require endocytosis within vesicles. Indeed our model proposes that, after binding to the cell surface, the protein is trapped within inverted lipid micelles that form within the plane of the membrane and reopen on the inner side of this membrane, releasing the protein into the cytoplasm. Secretion regulation can therefore occur at several levels: nuclear import, nuclear export, access to the secretory compartment and fusion of this compartment with the plasma membrane.

internalization as well as the formation of inverted micelles (Derossi et al. 1998).

Because inverted micelles do form when penetratin peptides are applied to vesicles composed of brain lipids, it was anticipated that the peptides and their attached cargoes translocate across the biological membranes within the hydrophilic pocket of the micelles (Derossi et al. 1998). This model, although not fully established, is very elegant because it explains most of the characteristics of this uncanny internalization, in particular the absence of a chiral receptor and an efficient transport at low temperature (no classical endocytosis permitted).

Internalized Homeodomains Regulate Transcription

As anticipated by the reader, if one wants to use homeoproteins or homeoprotein-derived peptides to modify the physiological properties of stem cells, or of other cell types, it is important that the internalized transcription factors be capable of regulating transcription. We are still working on the ability of full-length transcription factors to regulate gene activity after internalization and nuclear addressing, and the results are thus not yet available. However, the use of the homeodomain (DNA-binding domain) of the Engrailed homeoprotein has allowed us to identify, by gene trap and differential display, several putative homeoprotein target genes, among which are two bona fide regulators of cytoskeleton dynamic.

The gene trap protocol consists in the internalization of the homeodomain in 100% of ES cells (with a *lacZ*-containing gene trap vector) and in identifying the clones with an up- or down-regulation of *lacZ*. Using this approach we could demonstrate that the Bullous Pemphigoid Antigen-1 (BPAG-1) is a direct target of several homeoproteins (Mainguy et al. 1999, 2000). This is an interesting finding because BPAG-1 might be involved in some diseases ("blistering disease" and dystonia musculorum) and also because it confirms that many homeoprotein targets are regulators of the cytoskeleton.

In this context, it is significant that another gene identified by homeodomain internalization (and differential display) is MAP1B, a microtubule associated protein involved in axon growth and regeneration (Montesinos et al. 2001). The in vivo confirmation that the strategy used in such target "fishing" experiments leads to the identification of true target genes has reassured us that this strategy could be used to drive embryonic or adult stem cells and progenitors into the differentiation pathways of therapeutic value. Another obvious application is to associate functional genomics with homeoprotein internalization to identify all interacting genes and, possibly, new pharmacological targets in regenerative strategies.

Indeed, because the third helix of the homeodomain can be used to internalize various types of cargoes, the transcription factors or DNA-binding domains that could be addressed to the interior of live cells are not limited to homeoproteins. For example, and to restrict our speculation to DA cells, it is conceivable to internalize into live cells a mixture of Penetratin-Nurr1, Engrailed and Ptx3 and to verify if the three transcription factors are necessary and sufficient to transform a stem cell or progenitor, say from the SVZ, into a bona fide dopaminergic neuron (Daadi and Weiss 1999; Goridis and Rohrer 2002). Even though this proposal is still highly speculative, the fact that internalization occurs in 100% of the cells represents a real technical advantage. Another advantage is that, assuming that the cells are irreversibly programmed, this protein therapy does not involve viral vectors.

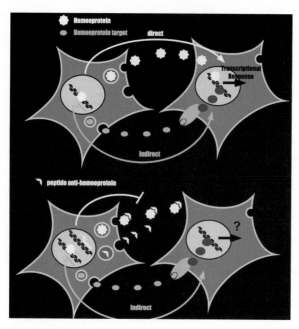

Fig. 2. Signaling with homeoproteins. The top part of this figure summarizes the two possible mechanisms of non-autonomous signal transduction. The classical (**indirect**) pathway involves the homeoprotein regulating the transcription of a target with signaling capacities, for example, a growth factor. This growth factor then gains access to a secretory compartment, is released in the extra-cellular milieu, binds a receptor and transduces a signal that leads to a transcriptional response of the receiving cell. Alternatively (**direct** pathway), it is proposed that some transcription factors, here homeoproteins, are secreted as described in Figure 1 and are internalized in abutting cells. Once internalized, they are addressed to the cell nucleus to regulate transcription. Indeed one cannot preclude other modes of signaling through the activation of second messengers or the regulation of mRNA translation. The bottom part of the figure illustrates the strategy that will be used to distinguish the respective functions of direct/indirect pathways. The idea is to design polypeptides or single chain antibodies that will be expressed by the cell and addressed into the secretory compartment. Through binding to the homeoprotein, the polypeptides should block or retard intercellular transfer. These peptides will also be used to modify the activity of the homeoprotein after internalization (using the penetratin strategy). They could thus define a new category of pharmacological agents with homeoprotein regulatory properties.

Why are Homeoproteins Transported?

Returning to the phenomenon of homeoprotein transport, the question of its developmental and physiological significance is not resolved. We have envisaged several hypotheses that will now be briefly summarized.

One hypothesis is that homeoprotein transfer is necessary to define territories and draw borders in the developing neuroepithelium. In most cases homeoprotein expression is initiated locally under the control of other transcription factors and then, in a second step, expands within the neuroepithelium. This secondary expansion is called homeogenic because it depends on the homeoprotein itself. Because most homeoproteins activate their own transcription, a simple and parsimonious model is that homeogenic expansion is due to the non-autonomous activation of its locus by the imported transcription factor. This infectious spreading of homeoprotein transcription will cease when invaded cells present a non-permissive context for the transcription of the "expanding" locus.

The latter, non-permissive context could be due to the expression of another transcription factor, for example another homeoprotein. Borders will thus form where the two transcription factors meet. This is, in fact, very often the case, as illustrated, for example, by Otx2 and Gbx2, Pax6 and Emx2, Pax6 and Gsh2 (Bishop et al. 2000; Martinez-Barbera et al. 2001; Toresson et al. 2000; Yun et al. 2000). In all cases, each homeoprotein activates its own transcription and represses that of its counterpart. A consequence of this model is that inhibiting the expression of one member of the couple will allow the expansion of the other, and this is the case. For example, decreasing the number of copies of Otx genes from four to one leads to a rostral expansion of Gbx2 and to a rostral displacement of the border materialized by the isthmus (Martinez-Barbera et al. 2001). Conversely, forcing the expression of Otx2 in the Gbx2 domain will caudalize the position of the border (and of the isthmus).

That homeoproteins activate their own synthesis in a non-autonomous manner might explain why regions in which the protein is present and not its mRNA (predicted by the non-autonomous hypothesis) were never observed. In addition, the model also predicts that non-autonomous activation responds to a very small number of imported transcription factors and thus that an amplifying mechanism exists. The nature of this mechanism is presently under study in the laboratory.

A second hypothesis is that homeoprotein transfer plays a role at later developmental stages and throughout adulthood. In this case we would like to see these proteins as true signaling molecules. We speculate that, in the developing axon, the import of a homeoprotein can serve as a guidance cue through the activation of second messengers or the regulation of translation. Indeed signaling could also be from the axon to the cells, with regulation taking place at three levels: second messengers, translation of localized mRNAs and transcription. This mode of signaling could also exist in stabilized synapses. In that case the transfer from the axon terminal to the post-synaptic site, with ensuing regulation of signaling, translation and transcription, would superpose topological information and neurotransmitter signaling. This is indeed of extreme interest in the functioning of a complex network: you not only hear the knocks at the door but you also know who is knocking.

Testing Hypotheses

There are several hints that some of the hypotheses proposed in the previous section are in part valid. First there is the demonstration that homeoproteins can travel between cells, are present in non-nuclear compartments, including axon terminals, and can be found in vivo within a secretory compartment (Joliot et al. 1997). The existence of highly conserved domains responsible for unconventional secretion and internalization is also interesting. Moreover, intercellular transport of homeoproteins has been demonstrated in plants even though, in plants, the existence of intracellular conducts (plasmodesmata) might explain the transfer (Lucas et al. 1995; Kim J. et al. 2002; Wu et al. 2002). Finally, as already mentioned, the competition between transcription factors for territorial control within the neuroepithelium fits with the hypothesis.

The best way to test the hypothesis would be to mutate one or several sequences responsible for homeoprotein transfer and to replace the wild-type gene by the mutated one in a knock-in experiment. The problem with this protocol is that the sequences important for transfer are within the homeodomain and are implied in its DNA-binding properties. Thus mutating these sequences would certainly have important autonomous consequences.

An alternative strategy being developed in our laboratory is to design polypeptides (e.g., single chain antibodies) that would bind the homeoproteins in the secretory compartment or in the extra-cellular space and block or retard their transfer. Using the phage display technology, we have identified several anti-Engrailed and anti-Otx2 polypeptides as well as an anti-Pax6 single chain antibody. Indeed, we also have the DNA sequences that code these polypeptides, and the next step is to use these minigenes to block homeoprotein transfer in vivo.

Conclusion

The observation that stem cells exist in the adult nervous system has opened new lines of investigation aimed at curing several diseases, in particular neurodegenerative diseases. We believe that the use of transcription factor cocktails will be one way to force the differentiation of these stem cells or progenitors into the desired developmental pathway (e.g., dopaminergic or cholinergic) and to target them into the appropriate region of the brain (e.g., the substantia nigra or the septum). This technology, which we could call protein therapy, differs from cell or gene therapies in the sense that it involves no viral vector and that, in the future, one can hope to program the cells in situ with non-invasive protocols. Finally, and to go beyond this practical view, the fact that large neuronal populations can be replaced throughout life sheds new light on the nature of neurodegenerative diseases. It is, indeed, tempting to propose that these diseases are not diseases of the aging cells but diseases of cell renewal, and thus developmental diseases of the adult. If so, it is the entire field of developmental biology that could become a part of adult physiology, a true revolution for our understanding of many still mysterious pathologies.

References

Alvarez-Buylla A, Theelen M, Nottebohm F (1988) Birth of projection neurons in the higher vocal center of the canary forebrain before, during, and after song learning. Proc. Natl. Acad. Sci. USA 85: 8722–8726

Bernier PJ, Bédard A, Vinet J, Lévesque M, Parent A (2002) Newly generated neurons in the amygdala and adjoining cortex of adult primates. Proc. Natl. Acad. Sci. USA 99: 11464-11469

Bishop KM, Goudreau G, O'Leary D (2000) Regulation of area identity in mammalian neocortex by Emx2 and Pax6. Science 288: 344–349

Clarke Dl, Johansson Cb, Wilbertz J, Veress B, Nilsson E, Karlström H, Lendahl U, Frisén J (2000) Generalized potential of adult neural stem cells. Science 288: 1660–1663

Daadi MM, Weiss S (1999) Generation of tyrosine hydroxylase-producing neurons from precursors of the embryonic and adult forebrain. J Neurosci 19: 4484–4497

Derossi D, Joliot AH, Chassaing G, Prochiantz A (1994) The third helix of Antennapedia homeodomain translocates through biological membranes. J Biol Chem 269: 10444–10450

Derossi D, Calvet S, Trembleau A, Brunissen A, Chassaing G, Prochiantz A (1996) Cell internalization of the third helix of the Antennapedia homeodomain is receptor-independent. J Biol Chem 271: 18188–18193

Derossi D, Chassaing G, Prochiantz A (1998) Trojan peptides: the penetratin system for intracellular delivery. Trends Cell Biol. 8: 84–87

Dupont E, Joliot A, Prochiantz A (2002) Penetratins. CRC Press

Goridis C, Rohrer H (2002) Specification of catecholaminergic and serotoninergic neurons. Nature Rev Neurosci 3: 531–541

Gould E, Reeves AJ, Graziano MSA, Gross CG (1999) Neurogenesis in the neocortex of adult primates. Science 286: 548–552

Horner PJ, Gage FH (2000) Regenerating the damaged central nervous system. Nature 407: 963–970

Joliot A, Trembleau A, Raposo G, Calvet S, Volovitch M, Prochiantz A (1997) Association of engrailed homeoproteins with vesicles presenting caveolae-like properties. Development 124: 1865–1875

Joliot A, Maizel A, Rosenberg D, Trembleau A, Dupas S, Volovitch M, Prochiantz A (1998) Identification of a signal sequence necessary for the unconventional secretion of Engrailed homeoprotein. Curr Biol 8: 856–863

Kim J-H, Auerbach JM, Rodriguez-Gomez JA, Velasco I, Gavin D, Lumelsky N, Lee S-H, Nguyen J, Sanchez-Pernaute R, Bankiewicz K, Mckay R (2002) Dopamine neurons derived from embryonic stem cells function in an animal model of Parkinson's disease. Nature 418: 50–56

Kim Jy, Yuan Z, Cilia M, Khalfan-Jagani Z, Jackson D (2002) Intercellular trafficking of a KNOTTED1 green fluorescent protein fusion in the leaf and shoot meristem of Arabidopsis. Proc Natl Acad Sci USA 99: 4103–4108.(NOT CITED IN TEXT)

Kornack DR, Rakic P (1999) Continuation of neurogenesis in the hippocampus of the adult macaque monkey. Proc Natl Acad Sci USA 96: 5768–5773.

Kornack DR, Rakic P (2001) The generation, migration, and differentiation of olfactory neurons in the adult primate brain. Proc Natl Acad Sci USA. 98: 4752–4757.

Kruger GM, Morrison SJ (2002) Brain repair by endogenous progenitors. Cell 110: 399–402

Kuhn HG, Svendsen CN (1999) Origins, functions and potential of adult neural stem cells. BioEssays 21: 625–630

Lois C, Alvarez-Buylla A (1993) Proliferating subventricular zone cells in the adult mammalian forebrain can differentiate into neurons and glia. Proc Natl Acad Sci USA 90: 2074–2077

Lucas WJ, Bouché-Pillon S, Jackson Dp, Nguyen L, Baker L, Ding B, Hake S (1995) Selective trafficking of KNOTTED1 homeodomain protein and its mRNA through plasmodesmata. Science 270: 1980–1983

Mainguy G, Erno H, Montesinos Ml, Lesaffre B, Wurst W, Volovitch M, Prochiantz A (1999) Regulation of epidermal bullous pemphigoid antigen 1 (BPAG1) synthesis by homeoprotein transcription factors. J Invest Dermatol 113: 643–650

Mainguy G, Luz Montesinos M, Lesaffre B, Zevnik B, Karasawa M, Kothary R, Wurst W, Prochiantz A, Volovitch M (2000) An induction gene trap for identifying a homeoprotein-regulated locus. Nature Biotechnol 18: 746–749

Maizel A, Bensaude O, Prochiantz A, Joliot A (1999) A short region of its homeodomain is necessary for Engrailed nuclear export and secretion. Development 126: 3183–3190

Maizel A, Tassetto M, Filhol O, Cochet C, Prochiantz A, Joliot A (2002) Engrailed homeoprotein secretion is a regulated process. Development 129: 3545–3553

Martinez-Barbera J, Signore M, Pilo Boyl P, Puelles E, Acampora D, Gogoi, Schubert F, Lumsden A, Simeone A (2001) Regionalisation of anterior neuroectoderm and its competence in responding to forebrain and midbrain inducing activities depends on mutual antagonism between OTX2 and GBX2. Development 128: 4789–4800

Montesinos MLl, Foucher I, Conradt M, Mainguy G, Robel L, Prochiantz A, Volovitch M (2001) The neuronal Microtubule Associated Protein 1 (MAP1B) is under homeoprotein transcriptional control. J Neurosci 21: 3350–3359

Munoz-Sanjuana I, Brivanlou AH (2002) Neural induction, the default model and embryonic stem cells. Nature Rev Neurosci 3: 271–280

Nottebohm F (2002) Neuronal replacement in adult brain. Brain Res Bull 57: 737–749

Prochiantz A (2000) Messenger proteins: homeoproteins, TAT and others. Curr Opin Cell Biol 12: 400–406

Shilo B-Z (2001) The organizer and beyond. Cell 106: 17–22

Toresson H, Potter Ss, Campbell K (2000) Genetic control of dorsal-ventral identity in the telencephalon: opposing roles for Pax6 and Gsh2. Development 127: 4361–4371

Wichterle H, Lieberam I, Porter JA, Jessell TM (2002) Directed differentiation of embryonic stem cells into motor neurons. Cell 110: 385–397

Wu X, Weigel D, Wigge PA (2002) Signaling in plants by intercellular RNA and protein movement. Genes Dev 16: 151–158

Yun K, Potter S, Rubenstein J (2000) Gsh2 and Pax6 play complementary roles in dorsoventral patterning of the mammalian telencephalon. Development 126: 193–205

Leukemia and Leukemic Stem Cells

C.H.M. Jamieson[1,2], E. Passegué[2] and I.L. Weissman[2]

Summary

Leukemias are cancers of the hematopoietic system. Like all cancers, several genetic and epigenetic events aid in the transition from normal to malignant cell. These usually, if not always, include at least: 1) avoidance of programmed cell death from intrinsic signals; 2) acquisition of poorly regulated or unregulated self-renewal capacity; 3) prevention of critical telomere shortening; 4) inhibition of differentiation to an increasing degree as the malignancy progresses; and 5) avoidance of innate and adaptive immune responses that cause the death and/or phagocytosis of tumor cells. Many of these processes are properties of the hematopoietic stem cell (HSC) and are highly regulated, yet the hallmark populations in the most advanced leukemias, e.g., the expanded population of leukemic blasts, are not HSC. The phenotypic identity of the leukemic stem cells (LSC), i.e., the only cells within the leukemia capable of propagating the disease, has not been clearly elucidated. In this speculative review, we have two goals: to discuss the stage of hematopoietic differentiation in which LSC reside, and to begin to understand how recurrent genetic changes, including translocations and inversions, and epigenetic changes, resulting in increased or decreased expression of selected genes, play roles in the above described behaviors of leukemia cells.

Introduction

Leukemia was first described by Virchow in 1856 in a monograph entitled "Die Leukemia," and literally translated means "white blood" (reviewed in Hoffman et al. 2000). Leukemia may be categorized as acute or chronic, lymphoid or myeloid and de novo or secondary to treatment or progression of myelodysplastic syndrome. Significant advances have been made in the treatment of both

---

[1] Division of Hematology, Department of Medicine, Stanford University School of Medicine, Stanford, California, USA

[2] Department of Pathology, Stanford University School of Medicine, Stanford, California, USA

Gage et al.
Stem Cells in the Nervous System:
Functional and Clinical Implications
© Springer-Verlag Berlin Heidelberg 2004

chronic and acute leukemias, including targeted small molecule inhibitor therapy (Gililand 2002: Stone 2002; O'Brien et al. 2003). In addition, allogeneic hematopoietic cell transplantation (HCT) for acute myelogenous (AML) and acute lymphoblastic leukemia (ALL), following myeloablative-conditioning regimens, affords cure rates of 50 to 75% (Blume and Forman 1992; Jamieson et al., submitted for publication). Many studies have focussed on the hallmark and most frequent cell types in leukemias, e.g., the leukemic blasts, including recent oligonucleotide and cDNA microarray analyses with attempts to provide a leukemic molecular signature (Golub et al. 1999). However, a growing body of evidence suggests that only a subset of cells in the leukemia hierarchy, the "leukemic stem cells", is responsible for propagating disease, although the exact phenotypic and functional characteristics of these LSC have yet to be clearly defined (Miyamoto et al. 1999; Bonnet and Dick 1997; Uchida et alo. 1998).

Hematopoietic Stem Cells and Leukemic Stem Cells

Hematopoietic Stem Cells (HSC)

The existence of HSC was first suggested in 1961 in experiments by Till, McCulloch and colleagues, who discovered that a population of clonogenic bone marrow cells, termed day 10 colony forming unit spleen (CFU-S_{10}), could give rise to myeloerythroid outcomes (Till and McCulloch 1961; Becker et al. 1963) and contained cells that reconstitute the hematolymphoid system subsequent to transplantation of lethally irradiated hosts (Wu et al. 1968). A subset of these colonies regenerated CFU-S_{10} (Till and McCulloch 1961), and a subset of these cells regenerated cells capable of hematolymphoid reconstitution. More recently, bone marrow fractionation experiments, using a collection of monoclonal antibodies and fluorescence-activated cell sorting (FACS), combined with the testing of these isolated populations in robust in vitro and in vivo assays of clonogenic hematopoietic precursors, have allowed researchers to define the phenotypic and functional characteristics of mouse HSC as well as of committed myeloid and lymphoid progenitors (Muller-Sieburg et al. 1986; Uchida and Weissman 1992; Morrison and Weissman 1994; Smith et al. 1991; Spangrude et al. 1988; Kondo et al. 1997; Akashi et al. 2000; Traver et al. 2000). These studies revealed that mouse HSC could be prospectively enriched or isolated using specific phenotypic marker combinations [c-Kit^{high}, Thy1.1^{low}, Lin^{neg}, Sca-1^{+}; KTLS], [Lin^{neg}, Sca-1^{+}, Rhodamine 123^{low}] (Muller-Sieburg et al. 1986; Uchida and Weissman 1992; Morrison and Weissman 1994; Smith et al. 1991; Spangrude et al. 1988; Kondo et al. 1997; Akashi et al. 2000; Traver et al. 2000; Manz et al. 2001; Na Nakorn et al. 2003; Weissman 2000; Morrison et al. 1995; Kondo et al. 2003; Uchida 1992; Wagers et al. 2003; Osawa et al. 1996). or [Lin^{neg}, CD34$^{-/int}$, c-Kit^{+}, Sca-1^{+}] (Osawa et al. 1996). Further research demonstrated that the KTLS HSC population could be subdivided into long-term HSC (LT-HSC), based on

Fig. 1. Mouse and human HSC have distinctive cell surface markers. Mouse HSC are c-kit$^+$, Thy-1^{lo}, Lin$^-$, Sca-1$^+$ (KTLS) whereas human HSC are CD90$^+$ (Thy-1$^+$), CD 34$^+$, CD38$^{-/lo}$, Lin$^-$.

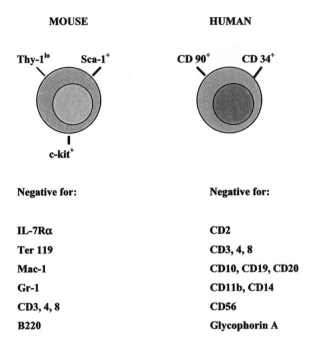

Negative for:

IL-7Rα

Ter 119

Mac-1

Gr-1

CD3, 4, 8

B220

Negative for:

CD2

CD3, 4, 8

CD10, CD19, CD20

CD11b, CD14

CD56

Glycophorin A

lack of cell surface Flk2 expression and low levels of Mac-1 expression, short-term HSC (ST-HSC), based on the acquisition of Flk2 expression and low levels of Mac-1, and finally, clonogenic multipotent progenitors (MPP), based on loss of Thy1.1 expression, low level expression of Mac-1 and CD4, and maintenance of cell surface Flk2 expression (Uchida and Weissman 1992; Christensen and Weissman 2001). Single LT-HSC transplanted in vivo with supporting host bone marrow could give long-term multilineage engraftment as well as expansion of the LT-HSC/ST-HSC/MPP pool (Morrison and Weissman 1994; Kondo et al. 2003; Wagers et al. 2002). Human HSC can also be highly enriched using unique phenotypic marker combinations [CD34$^+$, Thy1.1$^+$ (CD90$^+$), CD38^{low}, Lin$^-$], although the cell surface markers that specify commitment from human LT-HSC to ST-HSC followed by differentiation into MPP remain to be determined (reviewed in Bonnet and Dick 1997; Baum et al. 1992; Tsukamoto et al. 1995; Terstappen et al. 1991) (Fig. 1).

Hematopoiesis takes place through the step-wise differentiation of multipotent HSC to generate a hierarchy of progenitor populations with progressively restricted developmental potential, ultimately leading to the production of multiple lineages of mature effector cells. HSC differentiate along both the lymphoid lineage, giving rise to common lymphoid progenitors (CLP), which then differentiate into T-cells, B-cells, CD8α$^+$ and CD8α$^-$ dendritic cells (Kondo et al. 1997; Traver et al. 2000) and NK cells, and the myeloid lineage, giving rise to common myeloid progenitors (CMP) and then to megakaryocyte-erythroid progenitors (MEP), granulocyte-macrophage progenitors (GMP)

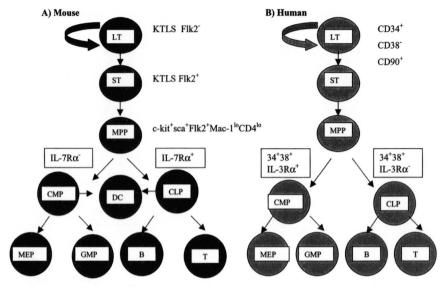

Fig. 2. A) Mouse HSC differentiation model. Long-term-HSC (LT) may self-renew or give rise to short-term-HSC (ST), which in turn differentiate into multipotent progenitors (MPP), followed by differentiation along the myeloid lineage, CMP (common myeloid progenitors; IL-7Rα-), or lymphoid lineage, CLP (common lymphoid progenitors; IL-7Rα+). CMP give rise to GMP (granulocyte-macrophage progenitors), MEP (megakaryocyte-erythroid progenitors), Mk (megakaryocyte progenitors) and CD8α-positive dendritic cells (DC) whereas CLP give rise to T, B and NK cell progenitors as well as CD8α-negative dendritic cells. **B)** Human HSC differentiation model. Human LT-HSC express CD34 and CD90 and are CD38- whereas the cell surface markers for ST-HSC and MPP remain to be defined. Human CMP are CD34+, CD38+ and IL-3Rα+ whereas human CLP are IL-3Rα-. Whether LSC share the same cell surface markers as normal HSC or more committed progenitors that have gained self-renewal capacity has yet to be determined.

and CD8α+ and CD8α- dendritic cells (Akashi et al. 2000; Traver et al. 2000; Manz et al. 2001; Fig. 2). While GMP appear to be the only source of monocytes, macrophages and neutrophils, another restricted megakaryocyte progenitor population (MkP) could contribute to megakaryocyte production (Na Nakorn et al. 2003).

Since the discovery of the phenotypic and functional properties of HSC and committed progenitors, the molecular mechanisms regulating self-renewal and lineage commitment within the hematopoietic system have become the focus of intensive investigation. Gene expression profiling of highly purified HSC and progenitor populations has yielded important insights into the transcriptional regulation of hematopoiesis (Ivanova et al. 2002; Terskikh et al. 2003; Terskikh et al. 2001; Park et al. 2002). A number of self-renewal genes such as Bmi-1, Hox, Notch and Wnt signaling pathway genes and their downstream targets have been found to play a critical role in maintaining and/or expanding the HSC pool (Park et al. 2003; Lessard and Sauvageau 2003; Antonchuk et al. 2002; Varnum-

Finney et al. 2000; Willert et al. 2002; Reya et al. 2003). The transcriptional profile of hematopoietic stem and progenitor cells at the single cell level has revealed that CLP and CMP express many genes operative in their progeny, more developmentally restricted progenitors (MEP and GMP for CMP; ProB and ProT for CLP) (Miyamoto et al. 2000, Kondo et al. 2000). In addition, specific transcription factors appear to be involved in commitment to the myeloid lineage (e.g., C/EBPα, PU.1, AML-1), to the erythroid lineage (e.g., GATA-2) and to the lymphoid lineages (e.g., Pax 5 and Ikaros) (reviewed in Graf 2002).

Leukemic Stem Cells (LSC)

Some models of leukemic pathogenesis suggest that large numbers of leukemic bone marrow cells are required to transplant the disease, suggesting that every cell within leukemic bone marrow is capable of proliferating and forming leukemia in transplanted recipients (reviewed in Reya et al. 2003). An alternative hypothesis is that leukemias contain rare cells with distinct phenotypic and functional characteristics, e.g., LSC, which are the only cells with the ability to self-renew extensively and give rise to leukemia in transplanted recipients. Like a normal stem cell, a LSC should be able to self-renew, home to the bone marrow, and have at least some progeny with high proliferative potential and some progeny with differentiation potential, thus explaining the heterogeneity of cell types within leukemia. A growing body of evidence supports the existence of cancer stem cells, including LSC, that may not necessarily overlap in phenotype with the stem cells of the tissue of origin (Miyamoto et al. 2000; Bonnet and Dick 1997; Park et al. 2003; Lessard and Sauvageau 2003; Lynch et al. 1972; Bruce and Gaag 1963; Wodinsky et al. 1967; Fialkow et al. 1977). Although a lot is known about transcriptional regulation of normal hematopoiesis, relatively little is known at the moment about the molecular switch(es) responsible for leukemic hematopoiesis.

The concept of a malignant stem cell arose from the finding that only 1 to 4% of transplanted lymphoma cells or lymphoma cell lines could form spleen colonies in recipient mice, although prospective isolations were not done and so poor plating efficiency could not be ruled out (Bruce and Gaag 1963; Wodinsky et al. 1967). In addition, G6PD isoenzyme studies in female patients with chronic myelogenous leukemia (CML) demonstrated a clonal pattern of G6PD isoforms in white blood cells, red blood cells and B cells, underscoring the possibility that there was an expansion of cells, preleukemic and/or CML, from a few clonal cells (Fialkow et al. 1977). The later discovery that over 95% of patients with CML harbored a characteristic chromosomal translocation - the Philadelphia chromosome, encoding the aberrant BCR-ABL fusion transcript, which was found to be present in both the primitive hematopoietic progenitors and their progeny -represented further evidence for the clonal etiology of at least one stage in the evolution of this type of leukemia (reviewed in Nowell and Hungerford 1960 and Sawyers 1999). Further studies demonstrated that BCR-ABL-positive HSC and progenitor cells share an autocrine IL-3/G-CSF mechanism that promotes survival in leukemic HSC and proliferation in more committed progenitors, with additional, still unknown, mutations in the regulatory functions of chronic phase primitive progenitors required for disease progression to blast crisis (Holyoake et al. 2001).

Additional evidence supporting the existence of LSC came from work performed in 1997, which demonstrated that a rare population (0.2% to 1%) of (CD34+, CD38-) cells from a number of AML patient bone marrow samples robustly transplanted leukemia in non-obese diabetic/severe combined (NOD/SCID) immunodeficient mice. These cells were referred to as SCID-leukemia initiating cells (SL-IC) because of their stem cell-like high proliferative potential, self-renewal capacity and ability to home to the marrow (Bonnet and Dick 1997). Furthermore, different studies demonstrated that the translocation between chromosomes 8 and 21, t(8;21), generates an AML1-ETO fusion transcription factor that can be detected in AML M2 leukemic blast cells but also in normal bone marrow cells, including HSC, obtained from AML patients in remission (Miyamoto et al. 1996, 2000; Miyoshi et al. 1991). The aberrant AML-1/ETO fusion protein inhibits transcriptional repression by the hematopoietic progenitor growth suppressing promyelocytic leukemia zinc finger (PLZF) protein (Melnick et al. 2000). AML-1/ETO also binds to and blocks C/EBP-α-mediated granulocytic differentiation (Westendorf et al. 1998) and inhibits myeloid differentiation by aberrantly recruiting histone-deacetylase activity (Puccetti et al. 2002). Interestingly, these prospectively isolated AML1-ETO-expressing HSC and their progeny are not leukemic and could differentiate into normal myelo-erythroid cells in vitro (Miyamoto et al. 2000). This observation suggests that the t(8;21) translocation occurs originally in normal HSC, and that additional mutations in a subset of these HSC or their progeny subsequently leads to leukemia. Since the original AML1-ETO-expressing normal HSC were [Lin-, CD34+, CD38-, CD90+] and the transplantable LSC were [Lin-, CD34+, CD38-, CD90-], the subsequent transforming mutation might have occurred either in downstream CD90 negative progenitors or in HSC that have lost CD90 expression as one of the first steps in leukemic pathogenesis (Miyoshi et al. 1991). These findings were underscored by the finding by Blair and co-workers that LSC from other human AML samples are also CD90 negative (Blair et al. 1997).

Although HSC are often the target of mutations or epigenetic events that are part of the progression to neoplastic transformation, committed progenitors may also become transformed. For example, in acute promyelocytic leukemia (APL), a translocation between chromosomes 15 and 17 gives rise to a fusion gene, PML/RARα, that is detectable by PCR in more committed progenitor cell populations but not in HSC-enriched cell populations (Turhan et al. 1995). This observation suggests that in APL, the transformation process may involve a more differentiated cell type than HSC and/or pluripotent progenitors that have been implicated in the other AML subtypes. In fact, mouse models of APL have been created by enforced expression of PML-RARα and bcl-2 in myeloid progenitors only (Grisolano et al. 1997; Kogan et al. 2001). Although leukemic transformation may occur in more committed progenitors, these populations may not function like normal committed progenitors in that they have reacquired characteristics, such as self-renewal, prolonged life span and telomere extension, that are normally properties of stem cells (Jamieson et al., manuscripts in preparation; Jaiswal et al. 2003; Cozzio et al., submitted for publication). Indeed, transformation of more differentiated cells has recently been described for breast cancer, where the cancer stem cell, as is typical of a HSC, has the capacity to serially transplant the same phenotype of cells, suggestive of self-renewal capacity (Al-Hajj et al. 2003). During normal hematopoiesis only HSC self-renew, whereas during leukemic hematopoiesis an accumulation of genetic changes directly in HSC or in the progeny of the HSC appears to lead to deregulated self-renewal or acquisition of self-renewing capability by cells that normally do not self-renew (Fig. 3).

Cell Autonomous Multi-Step Leukemic Pathogenesis

The multi-step model of carcinogenesis was originally postulated to require a clonal event causing uncontrolled proliferation that, together with mutations blocking cellular differentiation, caused malignant transformation (reviewed in Knudson 2001; Land et al. 1983). More recently, another class of proto-oncogene has been identified that either suppresses or promotes programmed cell death, or apoptosis, and plays critical roles in oncogenesis, as well as in regulating normal hematopoiesis (reviewed in Hawkins and Vaux 1997; Domen 2000). Although, genetic rearrangements and leukemia-associated fusion genes interfere with critical regulatory aspects of hematopoietic differentiation programs and thereby dictate the nature of the leukemia, they require additional cooperative mutations to induce full leukemic transformation. Transgenic mouse models have been instrumental in dissecting the aberrant molecular pathways leading to leukemia (Grisolano et al. 1997; Kogan et al. 2001; Jamieson et al., manuscripts in preparation; Jaiswal et al. 2003; Cozzio et al., submitted for publication; Hahn et al. 1999; Hanahan and Weinberg 2000). The development of leukemia is a multi-step process in which increasing numbers of mutations in cell autonomous pathways regulating self-renewal, proliferation, genomic stability and apoptosis as well as in non-autonomous pathways such as phagocytosis and immune surveillance, result in an increasingly transformed clonal population of leukemic cells such as LSC (Reya et al. 2001; Hanahan and Weinberg 2000).

PRELEUKEMIA

Events

Telomerase reverse transcriptase overexpression (**TERT***)

Fas deficiency (**Fas⁻**)

Fas⁻ TERT*

Genomic instability (**c-*myc***)

Resistance to apoptosis

(e.g., **Bcl-2** overexpression)

Fas⁻, Bcl-2*

Resistance to phagocytosis

(e.g., increased **CD47** expression)

LEUKEMIA

Acquisition of self-renewal capacity

(e.g., **β-catenin** overexpression)

Fas⁻ Bcl-2* CD47*

Fas⁻ Bcl-2* CD47*

β-catenin*

Genomic

instability

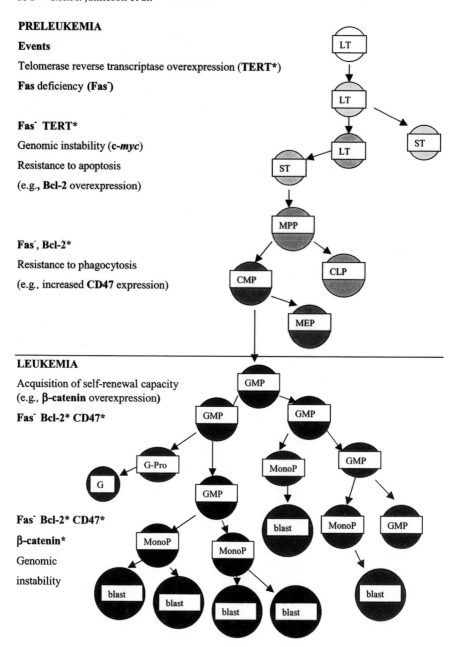

Fig. 3. A speculative view of AML development without cytogenetic abnormalities.

Resistance to Apoptosis

Regulation of programmed cell death or apoptosis is critical for normal tissue homeostasis. Deregulated apoptosis has been implicated in the pathogenesis of a number of hematological malignancies, including non-Hodgkin's lymphoma, acute myelogenous leukemia and chronic myelogenous leukemia. Mutations in specific molecular pathways controlling apoptosis, such as Bcl-2 and Fas deficiency, have been relatively well characterized in large part as a result of work performed with transgenic mouse models of hematological cancers.

Bcl-2 over-expression. The Bcl-2 family members, including Bcl-2, Bcl-x(L), Mcl-1 and A1, are anti-apoptotic proteins whereas their binding partners, such as Bax, Bad and Bak, promote apoptosis (reviewed in Delia et al. 1992). Thus, overexpression of Bcl-2 genes or underexpression of Bax family members would be expected to predispose toward cancer. Of note, Bcl-2 was first identified in follicular lymphoma upon characterization of a characteristic chromosomal translocation t(14;18) chromosomal translocation (Cleary et al. 1986). Although Bcl-2 has a well-established role in lymphomagenesis, deregulation of Bcl-2 gene expression also appears to be important in leukemic transformation of myeloid cells. The fusion protein AML1-ETO binds to the Bcl-2 promoter, resulting in increased Bcl-2 expression (Klampfer et al. 1996). In fact, cells from most types of human AML have been found to express Bcl-2 at much higher levels than their normal counterparts (Klampfer et al. 1996; Bensi et al. 1995). Together, these results suggest that activating mutations in Bcl-2 gene family members may be critical events in the multi-step process of leukemogenesis. Hence, the increased survival provided by enforced Bcl-2 gene expression may allow sufficient time for the acquisition of additional oncogenic mutations, a mechanism thought to underlie the transition from chronic to acute leukemia (reviewed in Rabbitts 1991). Consistently, enforced expression of Bcl-2 in all cells of the hematopoietic lineage, including HSCs, using the class I MHC promoter leads to increased HSC numbers and survival and to increased incidence of a variety of leukemias that develop late in life, thereby illustrating the necessity for additional mutations required for transformation (Domen et al. 1998). While the deregulation of Bcl-2 expression is found in many human cancers, overexpression of Bcl-2 in a transgenic mouse model has been found to be relatively benign in terms of cellular transformation (reviewed in Cory et al. 1994). Enforced expression of Bcl-2 in the myeloid lineage using the hMRP8 promoter (hMRP8^{Bcl-2}) leads to a disease that is similar to human chronic myelomonocytic leukemia (CMML) that is typified by progressive monocytosis, splenomegaly and neutropenia as the mice age (described in Traver et al. 1998). However, despite decreased survival compared to littermates, these mice rarely develop acute leukemia and they require additional mutations to promote AML, as has been shown for lymphoid leukemias (Vaux et al. 1988). Hence, myeloid-targeted overexpression of Bcl-2 greatly increases the incidence of CML-like disease in hMRP8^{BCR-}

[ABL]hMRP8[Bcl-2] double transgenic mice (Jaiswal et al. 2003), as well as the incidence of acute promyelocytic leukemia (APL) in hMRP8[PML/RARα] hMRP8Bcl-2 double transgenic mice (Kogan et al. 2001).

Fas deficiency. While deficiencies in Fas receptor (CD95) expression or other components of the Fas signaling pathway have previously been associated with autoimmunity (reviewed in Green and Ferguson 2001; Landowski et al. 2001), recent reports suggest that Fas mutations are important in the pathogenesis of a variety of human leukemias (Bouscary et al. 1997; Rumi et al. 1997). Granulocytes and myeloblasts are known to express high levels of Fas (Liles et al. 1996), and myeloid leukemic bone marrow blasts have been shown to have functional deficiencies in the Fas signaling pathway (Robertson et al. 1995). Interestingly, when transgenic mice with myeloid targeted overexpression of BCL-2 (hMRP8[Bcl-2]) are crossed with Fas-deficient (Fas[lpr/lpr]) mice, 15% of double transgenic (Fas[lpr/lpr]hMRP8[Bcl-2]) progeny develop AML before eight weeks post-partum (Traver et al. 1998). These results underscore the importance of the multiple steps required in leukemic pathogenesis, in that mutations in two distinct apoptotic pathways, Bcl-2 and Fas, are required, although additional mutations are still required for full leukemic transformation. Indeed, recent experiments demonstrate that leukemic transplantation potential is increased by 1) transplantation of leukemic cells into T cell deficient mice, 2) transplantation of leukemic GMP that are enriched for the LSC population, and 3) transplantation of leukemic progenitors with increased self-renewal capacity as a result of enforced expression of β-catenin (Jamieson et al., manuscripts in preparation; Jaiswal et al. 2003).

Taken together, these observations indicate that prevention of cell death is one of the critical events in myeloid leukemogenesis. Defects in apoptotic pathways could allow LSC to survive and be able to accumulate additional mutations, thereby promoting full leukemic transformation.

Self-Renewal

LSC require self-renewal capacity to propagate leukemia. Although the molecular mechanisms controlling HSC self-renewal activity have not been entirely elucidated, a number of genes have been shown to be critical for the maintenance of normal stem cell self-renewal, including Wnt, Bmi-1, Hox, Sonic hedgehog (Shh) and Notch pathway genes. Deregulation of a number of these genes has recently been implicated in the pathogenesis of a variety of cancers, including leukemia (reviewed in Reya et al. 2001, 2003; Park et al. 2003; Lessard and Sauvageau 2003; Antonchuk et al. 2002; Varnum-Finney et al. 2000; Willert et al. 2002).

Wnt. Wnts are highly hydrophobic, palmitoylated secreted signaling proteins that play a pivotal role in many phases of embryogenesis, including determination of cell fate, shape and polarity. Earlier work by Nusse Varmus and colleagues demonstrated that Wnt-3 is homologous to int-1/Wnt-1 and is activated by proviral insertion in mouse mammary tumors (MMTV). Transcriptional activation of int-1/Wnt-1 by proviral integration was found to induce mammary gland hyperplasia and adenocarcinomas in both male and female mice, suggesting that Wnt-1 is involved in initiating events in the multistep pathogenesis of cancer (reviewed in Nusse 2003; Nusse et al. 1984; Tsukamoto et al. 1988; Roelink et al. 1990). Recently, the different components of the Wnt signaling pathway have been elucidated (reviewed in Nusse 2003; Nusse et al. 1984; Tsukamoto et al. 1988; Roelink et al. 1990). Canonical Wnt signaling is usually activated by binding of Wnts to the seven transmembrane Frizzled receptors, resulting in ternary complex formation with LRP6 and activation of the Dishevelled (Dsh) gene product. Wnt signal transduction is tightly regulated at the cell surface by Dickkopf (Dkk), a secreted inhibitor that binds to the co-receptor, LDL-receptor related protein 6 (LRP6), thereby facilitating interaction with Kremen leading to Dkk-LRP6 complex internalization and inactivation (Rothbacher and Patrick 2002). Active Dsh inhibits the association of GSK3β and Axin with β-catenin, leading to reduced phosphorylation and nuclear translocation of β-catenin. Nuclear β-catenin associates with the LEF/TCF family of transcription factors and activates transcription of downstream target genes, including the proliferation associated genes cyclin D1 and c-myc (Rothbacher and Patrick 2002; Capelluto et al. 2002; Freemantle et al. 2002; Winn et al. 2002; Spiegelman et al. 2000; Shtutman et al. 2002; Austin et al. 1997; Van den Berg et al. 1998).

Overexpression of β-catenin, a critical downstream mediator of Wnt signaling and a transcriptional co-activator with LEF/TCF, has been implicated in the pathogenesis of colon, lung and breast cancers (Rothbacher and Patrick 2002; Capelluto et al. 2002). Although the role of elevated levels of β-catenin and subsequent LEF/TCF activation in the pathogenesis of leukemia has yet to be elucidated, β-catenin has been shown to play a pivotal role in HSC self-renewal and expansion (Willert et al. 2002; Reya et al. 2003). Retroviral transduction of constitutively activated β-catenin into HSC leads to their expansion with minimal differentiation in vitro, and retroviral transduction of the Wnt pathway inhibitor Axin leads to inhibition of HSC proliferation, increased HSC death in vitro and reduced reconstitution in vivo (Reya et al. 2003). Soluble partially purified Wnt proteins obtained from conditioned supernatants have also been shown to influence the proliferation of (CD34+) hematopoietic progenitors isolated from mouse fetal livers and human bone marrow (Willert et al. 2002; Shtutman et al. 2002; Van den Berg et al. 1998). Recently, purified Wnt3A has been shown to act on highly purified LT-HSCs to cause up to a 300-fold expansion of LT-HSCs identified both phenotypically and functionally (Reya et al. 2003). However, the molecular mechanisms by which Wnt signaling influences HSC self-renewal remain to be elucidated. In addition, the role of Wnts and β-catenin in LSC self-renewal and the pathogenesis of leukemia remains to be understood and thus is the focus of intensive research efforts.

HOX. Like Wnts, homeobox (HOX) transcription factors play a critical role in pattern specification and organogenesis during embryogenesis as well as adult tissue homeostasis (Shanmugam et al. 1999). HOX, PBX and MEIS bind DNA via a homeodomain. PBX transcription factors bind DNA cooperatively as heterodimers with MEIS and HOX transcription factors (Shanmugam et al. 1999). Members of the A, B, and C HOX gene clusters are expressed at high levels in primitive human progenitor populations (CD34+) with down-regulation during later stages of hematopoietic differentiation (Antonchuk et al. 2002; Schiedlmeier et al. 2003). Specific HOX genes such as HOXB4 have been shown to expand HSC both in vitro and in vivo in the absence of differentiation, suggesting that HOXB4 overexpression increases HSC self-renewal capacity, without increasing leukemogenic potential, but also impairs lympho-myeloid differentiation capacity (Antonchuk et al. 2002; Giannola et al. 2000; Schiedlmeier et al. 2003).). Interestingly, Wnt 3A-stimulated mouse HSC upregulate the expression of HOXB4, which might unite these pathways (Reya et al. 2003). Conversely, overexpression of HOXA9 and MEIS1 homeobox genes occurs frequently in human myeloid leukemias (Lawrence et al. 1996). Similarly, overexpression of HOXA10 alters murine myeloid and lymphoid differentiation, resulting in AML (Thorsteinsdottir et al. 1997). Finally, mice transplanted with bone marrow cells that are transduced with the human specific fusion gene, NUP98-HOXD13, together with the HOX cofactor, MEIS1, rapidly develop transplantable AML (Pienault et al. 2003). Thus, aberrant regulation of specific subsets of Hox genes together with their cofactors within LSC may play a role in the development of leukemia.

Notch. Activation of Notch by its ligand Jagged-1 in cultured HSCs also results in an increased amount of primitive progenitor activity, both in vitro and following transplantation in vivo, indicating that Notch activation promotes HSC self-renewal (Varnum-Finney et al. 2000; Karanu et al. 2000). As with HoxB4, Wnt3A- activated HSC upregulate expression of Notch-1 (Reya et al. 2003).

Sonic Hedgehog. Similar to Notch activation, primitive human progenitors [Lin⁻, CD34+, CD38⁻] demonstrate increased self-renewal capability in response to Shh stimulation in vitro, indicating that the Sonic hedgehog pathway may also play a role in regulating self-renewal of HSC, as in other types of tissue stem cells (Taipale and Beachy 2001; Bhardwaj et al. 2000), and may conceivably play a role in LSC self-renewal.

Taken together, these observations indicate that deregulating the pathways involved in controlling normal self-renewal in HSC may lead to leukemic transformation. It will be important to test whether LSC arising from committed progenitors also use the same pathways or novel gene expression pathways for their self-renewal.

Cell Proliferation and Differentiation

Like normal HSC, LSC undergo extensive proliferation. As HSC mature from the long-term, self-renewing pool into multipotent progenitors, they progressively lose their potential to self-renew but become more mitotically active. In young mice, the frequency of HSC in hematopoietic tissues is relatively constant (Morrison et al. 1995), and HSC have long been considered to be a resting cell population, with only a few stem cells contributing to steady-state hematopoiesis. In fact, recent studies have shown that, in young adult mice, about 8-10% of LT-HSC randomly enter the cell cycle per day, with all HSC entering the cell cycle in one to three months (Bradford et al. 1997; Cheshier et al. 1999). Although the rate at which human HSC cycle is currently unknown, assuming a comparable rate to that of the mouse would result in a very large number of cell divisions that an HSC can undergo (Vaziri et al. 1994).

Telomerase. Both tumorigenic cells and HSC share the capacity to avoid fatal telomere shortening through the action of the telomerase complex (Allsopp et al. 1995; Morrison et al. 1996). Mouse LT-HSC and malignant cells contain comparable telomerase activity whereas, ST-HSC and MPP have significantly lower telomerase activity (Allsopp et al. 2001). Enriched human stem/progenitor cell populations show telomere shortening with age (Allsopp et al. 1995), as do mouse LT-HSC that undergo many divisions during serial transplantation (Traver et al. 1998). These results are consistent with the hypothesis that telomere shortening may limit the replicative capacity of HSC. Serial transplantation has often been taken as a measure of the replicative life span of hematopoietic cells; however, regardless of telomere length, HSC bone marrow reconstitution capacity is severely reduced by the fifth serial transplant (Allsopp et al. 2003a,b). Recent experiments with serially transplanted LSCs isolated from a Fas[lpr/lpr]hMRP8[bcl2] mouse transgenic model of myeloid leukemia suggest that LSC are not reduced but rather expanded by the fifth serial transplant and thus, unlike normal HSC, appear to have an enhanced replicative life span (Jamieson et al., manuscript in preparation).

c-myc. The transcriptional regulator and proto-oncogene c-myc plays an important part in regulating cell cycle progression, metabolism, differentiation, cell adhesion, apoptosis, and hematopoietic homeostasis and in the pathogenesis of a number of malignancies, including leukemia, by interacting with a network of transcription factors (Hoffman et al. 2002; Langenau et al. 2003). A number of studies demonstrate the need for at least one or two mutations in addition to c-myc deregulation for full malignant transformation. Transgenic mice expressing a hematopoietic cell-inducible c-myc gene develop myeloid leukemia and lymphoma that regresses with subsequent transgene inactivation (Felsher and Bishop 1999), suggesting that c-myc plays a role in both the initiation and the maintenance of the tumors. Furthermore, when c-myc expression is deregulated in M1 myeloid leukemic cells and normal myeloid bone marrow cells, terminal differentiation is arrested, cell growth is inhibited and Fas-pathway regulated apoptosis is induced, suggesting that c-myc deregulation has pleiotropic effects (Hoffman et al. 2002). In the context of defective Fas-induced cell death, mutations in transcription factors (e.g., *c-fos, egr-1* and C/EBPα) that normally down-regulate *c-myc* expression during differentiation could enable LSC to survive and proliferate without differentiating, thereby contributing to leukemic pathogenesis.

BCR-ABL and other leukemia-associated fusion proteins. Leukemia-associated fusion proteins generally function as aberrantly activated signaling mechanisms or transcriptional regulators that directly interfere with the hematopoietic differentiation program (reviewed in Tenen et al. 1997; Alcalay et al. 2001; Tenen 2003). The 210 kD fusion protein product of BCR-ABL differs from the normal 145 kD c-*abl* protein product in its preferential localization within the cytoplasm and its constitutive tyrosine kinase activity. Both localization within the cell and constitutive tyrosine kinase activity play an important role in BCR-ABL-induced leukemic transformation (reviewed in Sawyers 1999). Localization of BCR-ABL within the cytoplasm activates c-*abl* and prevents its nuclear translocation, thereby preventing c- *abl* -mediated induction of apoptosis (Ling et al. 2003). Although the precise role of BCR-ABL in leukemic transformation has not been fully elucidated, a number of studies suggest that BCR-ABL alters motility, migration and adhesion of BCR-ABL-transformed cells, as well as increasing their resistance to apoptosis via BCR-ABL-driven overexpression of the anti-apoptotic proteins Bcl-x(L), A1 and Pim-1 (reviewed in –Wang 2000; Sattler et al. 2003; Schaller and Schaefer 2001; Nieborowska-Skorska et al. 2002). BCR-ABL also appears to deregulate BCR, resulting in oncogenic activation of c- *abl* and c-*myc* (Mahon et al. 2003). For the majority of AML-associated fusion proteins, one of the two components of each fusion protein is generally a transcription factor (AML-1, CBFβ, RARα) whereas the other partner is more variable in function but is often involved in the control of cell survival and apoptosis, such as the nuclear structure protein PML (reviewed in Tenen et al. 1997). Studies with a number of different experimental systems demonstrate that the AML-associated fusion proteins cause maturation arrest at specific stages of myeloid differentiation, depending upon the nature of the fusion protein expressed. Common mechanisms are utilized by these leukemia-associated fusion proteins, including recruitment of aberrant co-repressor complexes, alteration in chromatin remodelling and disruption in subnuclear compartments (reviewed in Alcalay et al. 2001; Tenen 2003).

Genomic Instability

Many cancers are genomically unstable, in part as a result of acquired defects in DNA repair mechanisms, and thus can readily acquire new mutations (Karlsson et al. 2003; Jain et al. 2003). If leukemias are maintained by a small subset of LSC, then genomic instability in these LSC could conceivably contribute to leukemic pathogenesis.

c-myc. A number of lines of evidence derived from the work performed on the c-*myc* oncogene indicate that during the initial stages of multi-step leukemic pathogenesis the leukemic genotype and phenotype are unstable and may regress unless additional chromosomal mutations are acquired (Felsher and Bishop 1999; Jain et al. 2002), but c-*myc* expression can result in the acquisition of genomic instability (Felsher and Bishop 1999). This result further emphasizes the fact that LSC require a number of mutations before being able to propagate the disease.

BCR-ABL. Although the aberrant product of the Philadelphia chromosome, BCR-ABL, is known to play a role in cell proliferation, differentiation, adhesion and apoptosis, recent work suggests that it also enhances repair of DNA lesions by facilitating homologous recombination repair but also by prolonging the time available for repairing DNA mutations through activation of G2/M cell cycle checkpoints as well as by increasing expression of anti-apoptotic proteins such as Bcl-x(L), A1 and Pim-1. However, because leukemias frequently have defective DNA repair mechanisms, the DNA mutations may not be repaired properly. Thus, aberrantly repaired DNA mutations may accumulate in BCR-ABL-transformed cells, resulting in genomic instability and progression of leukemia (reviewed in Skorski 2002).

Inherited DNA repair defects. A number of inherited defects in DNA repair are associated with an increased risk of leukemia, including Fanconi anemia (FA), Ataxia telangiectasia and Bloom syndrome. FA is an inherited bone marrow failure syndrome typified by skeletal defects as well as visceral and skin pigmentation abnormalities and a predisposition for developing myelodysplastic syndrome (MDS) and AML (D'Andrea and Grompe 1997). Eight complementation groups, including the BRCA2 (FANCD1) gene, have been implicated in Fanconi anemia. Defects in FA genes result in spontaneous chromosomal breakage, increased susceptibility to alkylating agent-induced DNA breaks and impaired ability to repair DNA damage (Lensch et al. 2003). Leukemic clones in children and adults with inherited FA protein dysfunction have complex cytogenetic abnormalities, including chromosomal loss and deletions, suggesting that cytogenetic instability within the LSC may contribute to leukemic pathogenesis (Lensch et al. 2003). Ataxia Telangiectasia (AT) patients with associated inactivation of the Ataxia Telangiectasia Mutated (ATM) gene appear to have an increased risk of developing B cell chronic lymphocytic leukemia (CLL) (Stankovic et al. 1999). However, the exact role of the ATM protein is currently unknown and its effects on predisposition to leukemia remain controversial. Nonetheless, defects in DNA repair and subsequent genomic instability in LSC could contribute to leukemic pathogenesis.

Cell Non-Autonomous Multi-Step Leukemic Pathogenesis

Although cell autonomous defects in LSC apoptotic pathways, self-renewal, proliferation and genomic stability appear to be central to the multi-step pathogenesis of leukemia, environmental factors, e.g., cell non-autonomous mechanisms, may provide the final steps in this multi-step process. Immunoregulatory deficits in the host, such as defects in phagocytosis and removal of LSC, impaired dendritic cell responses to LSC, as well as impaired development of mature T cells capable of recognizing and eliminating LSC, may contribute to disease progression.

Immune Surveillance

The hypothesis of immune surveillance has been strengthened in recent years because of the consistent finding that patients with graft-versus-host disease have lower relapse rates for leukemia, lymphoma, multiple myeloma and even renal cell carcinoma after myeloablative or non-myeloablative hematopoietic cell transplantation (HCT; Barrett 2003; Michalek et al. 2003; Maloney et al. 2002; Bregni et al. 2002; Boyer et al. 1997). In addition, some patients who have relapsed post-HCT attain a complete remission after the infusion of donor lymphocytes (DLI), suggesting that a graft-versus-tumor effect is critical in the maintenance of remission. Furthermore, patients who have been immunocompromised for protracted periods as a result of HCT, but more commonly following solid organ transplantation, may develop post-transplant lymphoproliferative disorder (PTLD). This malignant proliferation of lymphocytes frequently regresses when immunosuppression is withdrawn, and regression may be enhanced with the use of DLI, suggesting that the immune system can play a critical role in eliminating malignant cells (Durandy 2001). Finally, patients with inherited immunodeficiency states have higher rates of leukemia, suggesting that the immune system is key in eliminating malignant cells (Allsopp et al. 2001; Allsopp et al. 2003a). However, the component(s) of the immune system responsible for mediating tumor regression has yet to be elucidated, as have the molecular mechanisms responsible for controlling anti-tumor immune responses. Recent advances in immunology have provided important insights into the potential molecular mechanisms controlling thymic education and survival of immature lymphocytes such as the β-catenin-TCF pathway (Ionnidis et al. 2001; Gounari et al. 2001). In addition, genes regulating ectopic expression of transcripts in the thymus could play an important role in maintaining the balance between tolerance to and elimination of tumor cells (Anderson et al. 2002; Pitkanen and Paterson 2003; Liston et al. 2003).

CD47 Expression. Recent studies demonstrate that leukemic cells from a number of murine transgenic models of leukemia may evade phagocytosis and other immune-mediated methods of LSC removal by overexpressing CD47 (Oldenborg et al. 2001; Mateo et al. 2002). CD47, also known as integrin-associated protein (IAP), is a ubiquitously expressed cell surface glycoprotein that interacts with a number of integrins including $\alpha_v\beta_3$, $\alpha_{IIb}\beta_3$ and $\alpha_2\beta_1$ thereby modulating leukocyte adhesion, migration, phagocytosis, cell motility and platelet activation. CD47 also mediates several key immune regulatory processes. CD47 plays a critical role in host defense by modulating Fcγ and complement receptor-mediated phagocytic signals. The consequences of CD47 activation depend on the way the molecule is engaged, the surface receptors it interacts with and its conformation, as well as its membrane localization and the cell type that it is expressed on (reviewed in Oldenborg et al. 2001; Mateo et al. 2002). Recently, CD47 deficiency in NOD mice was shown to markedly accelerate the development of a lethal autoimmune hemolytic anemia in part because of decreased CD47 expression on the surface of red blood cells, resulting in increased erythrophagocytosis. However, mouse (CD34$^+$) progenitors appear to be relatively resistant to CD47-induced killing (Oldenborg et al. 2002; Vernon-Wilson et al. 2000). Recent work with mouse transgenic models of leukemia demonstrates a marked overexpression of CD47 by leukemic blasts, suggesting that CD47 may play a role in leukemic pathogenesis by allowing LSC and/or their progeny to escape phagocytosis (Jamieson et al. manuscripts in preparation; Jaiswal et al. 2003). Moreover, CD47 is a ligand for the rat macrophage membrane signal regulatory protein SIRP (OX41) and human SIRPalpha (Vernon-Wilson et al. 2000). CD47 also behaves as a marker of self by ligating the macrophage inhibitory receptor signal regulatory protein α (SIRPα), thus impeding macrophage-mediated phagocytosis primarily in SIRPα-expressing myeloid and neuronal cells (Mateo et al. 2002; Oldenborg et al. 2002). In addition, a number of investigators have reported that CD47 ligation decreases IL-12 production and hence inhibits maturation of dendritic cells as well as the maturation of naïve T cells into Th1 effector cells (Armant et al. 1999; Demeure et al. 2000; Avice et al. 2000; Latour et al. 2001). Thus, aberrant expression of CD47 by LSC and leukemic blasts may inhibit immune-mediated elimination of leukemogenic cells.

Conclusions

A compelling body of work argues for the existence of LSC, although their exact phenotypic and functional characteristics have yet to be elucidated. Whether LSC share the same phenotypic markers as normal HSC or represent more committed progenitors that have acquired stem cell characteristics – such as self-renewal and high proliferative capacity, ability to home to bone marrow and to undergo multi-lineage differentiation – is a focus of intensive investigation. However, in keeping with the concept of the multi-step pathogenesis of leukemia, a number of cell autonomous defects, including aberrantly increased self-renewal capacity, increased proliferative potential, genomic instability and resistance to apoptosis, as well as cell non-autonomous defects, including resistance to phagocytosis and impaired dendritic and T cell-mediated elimination of LSC, could enable LSC to gain a proliferative advantage over normal HSC and thereby facilitate leukemic progression. If LSC could be accurately identified based on characteristic phenotypic markers, they could be used to describe, for the first time, the gene expression profile of true leukemia cells rather than their non-malignant progeny. This approach would provide a list of target genes that, after validation, could be used to test new molecularly targeted drugs and also be used as targets of T and B cell immunity. Identification and elimination of LSC within leukemic marrow could conceivably remove one of the greatest barriers to curative hematopoietic cell transplantation – leukemic relapse.

Acknowledgements

C.H.M.J. is supported by a Yu-Bechmann fellowship for Genomics and Oncology at the Center for Clinical Immunology at Stanford (CCIS) and E.P. is a fellow of the Jose Carreras Foundation (FIJC-01/EDTHOMAS). This work was supported by the National Institutes of Health (CA55209 and CA86017) and a De Villier Award from the Leukemia and Lymphoma Society (ILW).

References

Akashi K, Traver D, Miyamoto T, Weissman IL (2000) A clonogenic common myeloid progenitor that give rise to all myeloid lineages. Nature 404:193–197

Alcalay M, Orleth A, Sebastiani C, Meani N, Chiaradonna F, Casciari C, Sciurpi MT, Gelmetti V, Riganelli D, Minucci S, Fagioli M, Pelicci PG (2001) Common themes in the pathogenesis of acute myeloid leukemia. Oncogene 20:5680–5694

Al-Hajj M, Wicha MS, Benito-Hernandez A, Morrison SJ, Clarke MF (2003) Prospective identification of tumorigenic breast cancer cells. Proc Natl Acad Sci USA 100: 3983–3988

Allsopp RC, Cheshier S, Weissman IL (2001) Telomere shortening accompanies increased cell cycle activity during serial transplantation of hematopoietic stem cells. J Exp Med 193: 917–924

Allsopp RC, Morin GB, Horner JW, DePinho R., Harley CB, Weissman IL (2003a) Effect of TERT over-expression on the long-term transplantation capacity of hematopoietic stem cells. Nature Med 9: 369–371

Allsopp RC, Morin GB, DePinho R, Harley CB, Weissman IL (2003b) Telomerase is required to slow telomere shortening and extend replicative lifespan of HSC during serial transplantation. Blood 102: 517–520

Allsopp RC, Chang E, Kashefi-Aazam M, Rogaev EI, Piatyszek MA, Shay JW, Harley CB (1995) Telomere shortening is associated with cell division in vitro and in vivo. Exp Cell Res 220: 194–200

Alter BP (2003) Cancer in Fanconi Anemia. Cancer 97:425–440

Anderson MS, Venanzi ES, Klein L, Chen Z, Berzins SP, Turley SJ, von Boehmer H, Bronson R, Dierich A, Benoist C, Mathis D (2002) Projection of an immunological self shadow within the thymus by the Aire protein. Science 298:1395–1401

Antonchuk J, Sauvageau G, Humphries RK (2002) Hox B4-induced expansion of adult hematopoietic stem cells ex vivo. Cell 109:39–45

Armant M, Avice MN, Hermann P, Rubio M, Kiniwa M, Delespesse G, Sarfati M (1999) CD47 ligation selectively downregulates human interleukin 12 production. J Exp Med 190:1175–1182

Austin TW, Solar GP, Ziegler FC, Liem L, Matthews W (1997) A role for the Wnt gene family in hematopoiesis: expansion of multilineage progenitor cells. Blood 89:3624–3635

Avice MN, Rubio M, Sergerie M, Delespesse G, Sarfati M (2000) CD47 ligation selectively inhibits the development of human naïve T cells into Th1 effectors. J Immunol 165:4624–4631

Barrett J (2003) Allogeneic stem cell transplantation for chronic myeloid leukemia. Semin Hematol 40:59–71

Baum CM, Weissman IL, Tsukamoto AS, Buckle AM, Peault B (1992) Isolation of a candidate human hematopoietic stem-cell population. Proc Natl Acad Sci USA 89:2804

Becker A, McCulloch E, Till J (1963) Cytologic demonstration of the clonal nature of spleen colonies derived from transplanted mouse marrow cells. Nature 197:452–454

Bensi L, Longo R, Vecchia A, Messora C, Garagnani L, Bernardi S, Tamassia MG, Sacchi S (1995) Bcl-2 oncoprotein expression in acute myeloid leukemia. Haematologica 80: 98–102

Bhardwaj G, Murdoch B, Wu D, Baker DP, Williams KP, Chadwick K, Ling, LE, Karanu, FN, Bhatia M (2000) Sonic hedgehog induces the proliferation of primitive human hematopoietic cells via BMP regulation. Nature Immunol 2: 172–180

Blair A, Hogge DE, Ailles LE, Lansdorp, PM, Sutherland HJ (1997) Lack of expression of Thy-1 (CD90) on acute myeloid leukemia cells with long-term proliferative ability in vitro and in vivo. Blood 89:3104–3112

Blume KG, Forman SJ. (1992) High-Dose etoposide (VP-16)-containing preparatory regimens in allogeneic and autologous bone marrow transplantation for hematologic malignancies. Semin Oncol 19 Suppl 13:63–66

Bonnet D, Dick JE (1997) Human acute myeloid leukemia is organized as a hierarchy that originates from a primitive hematopoietic stem cell. Nature Med 3:730–737

Bouscary D, De Vos J, Guesnu M, Jondeau K, Viguier F, Melle J, Picard F, Dreyfus F, Fontenay-Roupie M (1997) Fas/Apo-1 (CD95) expression and apoptosis in patients with myelodysplastic syndromes. Leukemia 11:839–845

Boyer MW, Vallera DA, Taylor PA, Gray GS, Katsanis E, Gorden K, Orchard PJ, Blazar BR (1997) The role of B7 costimulation by murine acute myeloid leukemia in the generation and function of a CD8+ T cell line with potent in vivo graft-versus-leukemia properties. Blood 89:3477–3485

Bradford GB, Williams B, Rossi R, Bertoncello I (1997) Quiescence, cycling, and turnover in the primitive hematopoietic stem cell compartment. Exp Hematol 25:445–453

Bregni M, Dodero A, Peccatori J, Pescarollo A, Bernardi M, Sassi I, Voena C, Zaniboni A, Bordignon C, Corradini P (2002) Nonmyeloablative conditioning followed by hematopoietic cell allografting and donor lymphocyte infusions for patients with metastatic renal and breast cancer. Blood 99:4234–4236

Bruce W, Gaag, H (1963) A quantitative assay for the number of murine lymphoma cells capable of proliferation in vivo. Nature 199: 79–80

Capelluto DGS, Kutateladze TG, Habas R, Finkielstein CV, He X, Overduin M (2002) The DIX domain targets disheveled to actin stress fibers and vesicular membranes. Nature 419: 726–729

Cheshier, SH, Morrison SJ, Liao X, Weissman, IL (1999) In vivo proliferation and cell cycle kinetics of long-term hematopoietic stem cells. Proc Natl Acad Sci USA 96:3120–3125

Christensen JL, Weissman IL (2001) Flk-2 is a marker in hematopoietic stem cell differentiation: a simple method to isolate long-term stem cells. Proc Natl Acad Sci USA 98:14541–14546

Cleary ML, Smith SD, Skar J (1986) Cloning and structural analysis of cDNAs for bcl-2 and a hybrid bcl-2/immunoglobulin transcript resulting from the t(14;18) translocation. Cell 10:19–28

Cory S, Harris AW, Strasser A (1994) Insights from transgenic mice regarding the role of bcl-2 in normal and neoplastic lymphoid cells. Philos Trans Royal Soc Lond B Biol Sci 345:289–295

D'Andrea AD and Grompe M (1997) Molecular biology of Fanconi anemia: implications for diagnosis and therapy. Blood 90:1725–1736

Delia D, Aiello A, Soligo D, Fontanella E, Melani C, Pezzella F, Pierotti MA, Della Porta G (1992) Bcl-2 proto-oncogene expression in normal and neoplastic human myeloid cells. Blood 79: 1291–1298

Demeure CE, Tanaka H, Mateo V, Rubio M, Delespesse G, Sarfati M (2000) CD47 engagement inhibits cytokine production and maturation of human dendritic cells. J Immunol 164: 2193–2199

Domen J (2000) The role of apoptosis in regulating hematopoiesis and hematopoietic stem cells. Immunol Res 22: 83–94

Domen, J, Gandy KL, Weissman IL (1998) Systemic overexpression of BCL-2 in the hematopoietic system protects transgenic mice from the consequences of lethal irradiation. Blood 91:2272–2282

Durandy A (2001) Anti-B cell and anti-cytokine therapy for the treatment of post-transplant lymphoproliferative disorder: past, present and future. Transpl Infect Dis 3:104–107

Felsher DW, Bishop JM (1999) Transient expression of MYC activity can elicit genomic instability and tumorigenesis. Proc Natl Acad Sci USA 96:3940–3944

Fialkow PJ, Jacobson RJ, Papayannopoulou T (1977) Chronic myelocytic leukemia: clonal origin in a stem cell common to the granulocyte, erythrocyte, platelet and monocyte/macrophage. Am J Med 63: 125–130

Freemantle SJ, Portland HB, Ewings K, Dmitrovsky F, DiPetrillo K, Sinella MJ, Dmitrovsky E (2002) Characterization and tissue-specific expression of human GSK-3-binding protein FRAT1 and FRAT2. Gene 292:17–27

Giannola DM, Shlomchik WD, Jegathesan M, Liebowitz D, Abrams CS, Kadesch T, Dancis A, Emerson SG (2000) Hematopoietic expression of HOXB4 is regulated in normal and leukemic stem cells through transcriptional activation of the HOXB4 promoter by upstream stimulating factor (USF)-1 and USF-2. J Exp Med 192:1479–1490

Gilliland DG (2002) Molecular genetics of human leukemias: New insights into therapy. Semin Hematol 39:6–11

Golub TR, Slonim DK, Tamayo P, Huard C, Gaasenbeek M, Mesirov JP, Coller H, Loh ML, Downing JR, Caligiuri MA, Bloomfield CD, Lander ES (1999) Molecular classification of cancer: class discovery and class prediction by gene expression monitoring. Science 286, 531–537

Gounari F, Aifantis I, Khazaie K, Hoeflinger S, Harada N, Taketo MM, von Boehmer H (2001) Somatic activation of β-catenin bypasses pre-TCR signalling and TCR selection in thymocyte development. Nature Immunol 2:863–869

Graf T (2002) Differentiation plasticity of hematopoietic cells. Blood 99:3089–3101

Green DR and Ferguson TA (2001) The role of Fas ligand in immune privilege. Nature Rev Mol Cell Biol 2: 917–924

Grisolano JL, Wesselschmidt RL, Pelicci PG, Ley TJ (1997) Altered myeloid development and acute leukemia in transgenic mice expressing PML-RAR alpha under control of Cathepsin G regulatory sequences. Blood 89:376–387

Hahn WC, Counter CM, Lundberg AS, Beijersbergen RL, Brooks MW, Weinberg, RA (1999) Creation of human tumor cells with defined genetic elements. Nature 400: 464–468

Hanahan D, Weinberg, RA (2000) The hallmarks of cancer. Cell 100:57–70

Hawkins CJ, Vaux, DL (1997) The role of the Bcl-2 family of apoptosis regulatory proteins in the immune system. Semin Immunol 9: 25–33

Hoffman B, Amanullah A, Shafarenko M, Liebermann DA (2002) The proto-oncogene c-myc in hematopoietic development and leukemogenesis. Oncogene 21:3414–3421

Hoffman R, Benz Jr E, Sanford J, Shattil SJ, Furie B, Cohen HJ, Silberstein LE, McGlave P, Strauss M (2000) Clinical manifestations of acute myeloid leukemia. In: Hoffman R, Strauss M (eds) Hematology: basic principles and practice. Churchill Livingstone, Philadelphia p. 999

Holyoake TL, Jiang X, Jorgensen HG, Graham S, Alcorn MJ, Laird C, Eaves AC, Eaves CJ (2001) Primitive quiescent leukemic cells from patients with chronic myeloid leukemia spontaneously initiate factor-independent growth in vitro in association with up-regulation of expression of interleukin-3. Blood 97:720–772

Ionnidis V, Beermann F, Clevers H, Held W (2001) The β-catenin-TCF-1 pathway ensures CD4+CD8+ thymocyte survival. Nature Immunol 2:691–697

Ivanova NB, Dimos JT, Schaniel C, Hackney JA, Moore KA, Lemischka IR (2002) A stem cell molecular signature. Science 298: 601–604

Jain M, Arvanitis C, Chu K, Dewey W, Leonhardt E, Trinh M, Sundberg CD, Bishop JM, Felsher DW (2002) Sustained loss of a neoplastic phenotype by brief inactivation of MYC. Science 297: 102–104

Jaiswal S, Traver D, Miyamoto T, Akashi K, Lagasse E, Weissman IL (2003) Expression of BCR/ABL and BCL-2 in myeloid progenitors leads to myeloid leukemias. Proc Natl Acad Sci USA 100: 10002–10007

Karanu FN, Murdoch B., Gallacher L, Wu DM, Koremoto M, Sakano S, Bhatia M (2000) The Notch ligand Jagged-1 represents a novel growth factor of human hematopoietic stem cells. J Exp Med 192:1365–1372

Karlsson A, Giuriato S, Tang F, Fung-Weier J, Levan G, Felsher DW (2003) Genomically complex lymphomas undergo sustained tumor regression upon MYC inactivation unless they acquire novel chromosomal translocations. Blood 101: 2797–2803

Klampfer L, Zhang, J, Zelenetz AO, Uchida H, Nimer SD (1996) The AML1/ETO fusion protein activates transcription of BCL-2. Proc Natl Acad Sci USA 93:14059–14064

Knudson AG (2001) Two genetic hits (more or less) to cancer. Nature Rev Cancer 1:157–162

Kogan SC, Brown DE, Shultz DB, Truong BT, Lallemand-Breitenbach V, Guillemin MC, Lagasse E, Weissman IL, Bishop JM (2001) BCL-2 cooperates with promyelocytic leukemia retinoic acid receptor alpha chimeric protein (PML-RARalpha) to block neutrophil differentiation and initiate acute leukemia. J Exp Med 193: 531–543

Kondo M, Weissman, IL, Akashi K (1997) Identification of clonogenic common lymphoid progenitors in mouse bone marrow. Cell 91:661–672

Kondo M, Scherer DC, Miyamoto T et al. (2000) Cell-fate conversion of lymphoid committed progenitors by instructive actions of cytokines. Nature 407:383–386

Kondo M, Wagers AJ, Manz MG, Prohaska SS, Scherer DC, Beilhack GF, Shizuru JA, Weissman IL (2003) Biology of hematopoietic stem cells and progenitors: Implications for clinical application. Annu Rev Immunol 21:759–806

Land H, Parada LF, Weinberg RA (1983) Cellular oncogenes and multistep carcinogenesis. Science 222:771–778

Landowski TH, Moscinski L, Burke R, Buyuksal I, Painter JS, Goldstein S, Dalton WS (2001) CD95 antigen mutations in hematopoietic malignancies. Leuk Lymphoma 42:835–846

Langenau DM, Traver D, Ferrando AA, Kutok JL, Aster JC, Kanki JP, Lin S, Prochownik E, Trede NS, Zon LI, Look AT (2003) Myc-induced T cell leukemia in transgenic zebrafish. Science 299: 887–890

Latour S, Tanaka H, Demeure C, Mateo V, Rubio M, Brown EJ, Maliszewski C, Lindberg FP, Oldenborg A, Ullrich A, Delespesse G, Sarfati M (2001) Bidirectional negative regulation of human T and dendritic cells by CD47 and its cognate receptor signal-regulator protein-alpha: down-regulation of IL-12 responsiveness and inhibition of dendritic cell activation. J Immunol 167:2547–2554

Lawrence HJ, Sauvageau G, Humphries RK, Largman C (1996) The role of HOX homeobox genes in normal and leukemic hematopoiesis. Stem Cells 14:281–291

Lensch MW, Tischkowitz M, Christianson TA, Reifsteck CA, Speckhart SA, Jakobs PM, O'Dwyer ME, Olson SB, Le Beau MM, Hodgson SV, Mathew CG, Larson RA, Bagby GC (2003) Acquired FANCA dysfunction and cytogenetic instability in adult acute myelogenous leukemia. Blood 102:7–16

Lessard J, Sauvageau G (2003) Bmi-1 determines the proliferative capacity of normal and leukaemic stem cells. Nature 423: 255–260

Liles WC, Kiener PA, Ledbetter JA, Aruffo A, Klebanoff SJ (1996) Differential expression of Fas (CD95) and Fas ligand on normal human phagocytes: implications for the regulation of apoptosis in neutrophils. J Exp Med 184:429–440

Ling X, Ma G, Sun T, Liu J, Arlinghaus RB (2003) Bcr and Abl interaction: oncogenic activation of c-Abl by sequestering Bcr. Cancer Res 63:298–303

Liston A, Lesage S, Wilson J, Peltonen L, Goodnow CC (2003) Aire regulates negative selection of organ-specific T cells. Nature Immunol 4:350–354

Lynch RG, Graff RJ, Sirisinha S, Simms ES, Eisen HN (1972) Myeloma proteins as tumor-specific transplantation antigens. Proc Natl Acad Sci USA 69:1540–1544

Mahon GM, Wang Y, Korus M, Kostenko E, Cheng L, Sun T, Arlinghaus RB, Whitehead IP (2003) The c-Myc oncoprotein interacts with Bcr. Curr Biol 13:437–441

Maloney DG, Sandmaier BM, Mackinnon S, Shizuru JA (2002) Non-myeloablative transplantation. Hematology (Am Soc Hematol Educ Program):392–421

Manz MG, Traver D, Miyamoto T, Weissman IL, Akashi K (2001) Dendritic cell potentials of early lymphoid and myeloid progenitors. Blood 97:3333–3341

Mateo V, Brown EJ, Biron G, Rubio M, Fischer A, Le Deist F, Sarfati M (2002) CD47-induced caspase-independent cell death in normal and leukemic cells: link between phosphatidylserine exposure and cytoskeleton organization. Blood 100:2882–2990

Melnick A, Carlile GW, McConnell MJ, Polinger A, Hiebert SW, Licht JD (2000) AML-1/ETO fusion protein is a dominant negative inhibitor of transcriptional repression by the promyelocytic leukemia zinc finger protein. Blood 96:3939–3947

Michalek J, Collins RH, Durrani HP, Vaclavkova P, Ruff LE, Douek DC, Vitetta ES (2003) Definitive separation of graft-versus-leukemia and graft-versus-host specific CD4+ T cells by virtue of their receptor beta loci sequences. Proc Natl Acad Sci USA 100:1180–1184

Miyamoto T, Nagafuji K, Akashi K, Harada M, Kyo T, Akashi T, Takenaka K, Mizuno S, Gondo H, Okamura T, Dohy H, Niho Y (1996) Persistence of multipotent progenitors expressing AML1-ETO transcripts in long-term remission patients with t(8;21) acute myelogenous leukemia. Blood 87: 4789–4796

Miyamoto T, Weissman IL, Akashi K (2000) AML1/ETO-expressing non-leukemic stem cells in acute myelogenous leukemia with 8;21 chromosomal translocation. Proc Natl Acad Sci USA 97:7521–7526

Miyamoto T, Iwasaki H, Reizis B, Ye M, Graf T, Weissman IL, Akashi K (2002) Myeloid or lymphoid promiscuity as a critical step in hematopoietic lineage commitment. Dev Cell 3: 137–147

Miyoshi H, Shimizu K, Kozu T, Maseki N, Kaneko Y, Ohki M (1991) t(8;21) breakpoints on chromosome 21 in acute myeloid leukemia are clustered within a limited region of a single gene, AML1. Proc Natl Acad Sci USA 88:10431–10434

Morrison SJ, Weissman IL (1994) The long-term repopulating subset of hematopoietic stem cells is deterministic and isolatable by phenotype. Immunity 1:661–673.

Morrison SJ, Uchida N, Weissman IL (1995) The biology of hematopoietic stem cells. Ann Rev Cell Dev Biol 11:35–71

Morrison SJ, Prowse KR, Ho P, Weissman IL (1996) Telomerase activity in hematopoietic cells is associated with self-renewal potential. Immunity 5:207–216.

Muller-Sieburg CF, Whitlock CA, Weissman IL (1986) Isolation of two early B lymphocyte progenitors from mouse marrow: a committed pre-pre-B cell and a clonogenic Thy-1-lo hematopoietic stem cell. Cell 44:653–662

Na Nakorn T, Miyamoto T, Weissman IL (2003) Characterization of mouse clonogenic megakaryocyte progenitors. Proc Natl Acad Sci USA 100:205–210

Nieborowska-Skorska M, Hoser G, Kossev P, Wasik MA, Skorski T (2002) Complementary functions of the antiapoptotic protein A1 and serine/threonine kinase pim-1 in the BCR/ABL-mediated leukemogenesis. Blood 99:4531–4539

Nowell PC, Hungerford DA (1960) A minute chromosome in human chronic granulocytic leukemia. Science 132:1497

Nusse, R (2003) Making head or tail of Dickkopf. Nature 411:255–256

Nusse R, van Ooyen A, Cox D, Fung YK, Varumus H (1984) Mode of proviral integration of a putative mammary oncogene (int-1) on mouse chromosome 15. Nature 307:131–136

O'Brien SG, Guilhot F, Larson RA, Gathmann I, Baccarani M, Cervantes F, Cornelissen JJ, Fischer T, Hochhaus A, Hughes T, Lechner K, Nielsen JL, Rousselot P, Reiffers J, Saglio G, Shepherd J, Simonsson B, Gratwohl A, Goldman JM, Kantarjian H, Taylor K, Verhoef G, Bolton AE, Capdeville R, Druker BJ (2003) IRIS Investigators. Imatinib compared with interferon and low-dose cytarabine for newly diagnosed chronic-phase chronic myeloid leukemia Imatinib compared with interferon and low-dose cytarabine for newly diagnosed chronic-phase chronic myeloid leukemia. New Engl J Med 348:994–100

Oldenborg PA, Gresham HD, Lindberg FP (2001) CD47-signal regulatory protein α (SIRPα) regulates Fcγ and complement receptor-mediated phagocytosis. J Exp Med 193:855–862

Oldenborg P-A, Gresham HD, Chen Y, Izui S, Lindberg FP (2002) Lethal autoimmune hemolytic anemia in CD47-deficient diabetic (NOD) mice. Blood 99:3500–3504

Osawa M, Hanada K, Hanada H, Nakauchi H (1996) Long-term lymphohematopoietic reconstitution by a single CD34-low/negative hematopoietic stem cell. Science 273:242–245

Park IK, He Y, Lin F et al. (2002) Differential gene expression profiling of adult murine hematopoietic stem cells. Blood 99:488–498

Park IK, Qian D, Kiel M, Becker MW, Pihalja M, Weissman IL, Morrison SJ, Clarke MF (2003) Bmi-1 is required for maintenance of adult self-renewing haematopoietic stem cells. Nature 423:302–305

Pineault N, Buske C, Feuring-Buske M, Abramovich C, Rosten P, Hogge DE, Aplan PD, Humphries RK (2003) Induction of acute myeloid leukemia (AML) in mice by the human leukemia-specific fusion gene NUP98-HOXD13 in concert with Meis1. Blood 101:4529–4538

Pitkanen J, Peterson P (2003) Autoimmune regulator: from loss of function to autoimmunity. Genes Immunol 4:12–21

Puccetti E, Obradovic D, Beissert T, Bianchini A, Washburn B, Chiaradonna F, Boehrer S, Hoelzer D, Ottmann OG, Pelicci PG, Nervi C, Ruthardt M (2002) AML-associated translocation products block vitamin D(3)-induced differentiation by sequestering the vitamin D(3) receptor. Cancer Res 62:7050–7058

Rabbitts TH (1991) Translocations, master genes, and differences between the origins of acute and chronic leukemias. Cell 67:641–644

Reya T, Morrison SJ, Clarke MF, Weissman IL (2001) Stem cells, cancer, and cancer stem cells. Nature 414:105–111

Reya T, Duncan AW, Ailles L, Domen J, Scherer DC, Willert K, Hintz L, Nusse R, Weissman IL (2003) A role for Wnt signalling in self-renewal of haematopoietic stem cells. Nature, 423:409–414

Robertson MJ, Manley TJ, Pichert G, Cameron C, Cochran KJ, Levine H, Ritz J (1995) Functional consequences of APO-1/Fas (CD95) antigen expression by normal and neoplastic hematopoietic cells. Leuk Lymphoma 17:51–61

Roelink H, Wagenaar E, Lopes da Silav S, Nusse R (1990) Wnt-3, a gene activated by proviral integration in mouse mammary tumors, is homologous to int-1/Wnt-1 and is normally expressed in mouse embryos and adult brain. Proc Natl Acad Sci USA 87:4519–4523.

Rothbacher Ute and Lemaire Patrick (2002) Crème de la Kremen of Wnt signaling inhibition. Nature Cell Biol 4:172

Rumi C, Rutella S, Lucia B.M, Teofili L, De Stefano V, Leone G (1997) APO-1/Fas receptor (CD95) is non-functionally expressed in acute promyelocytic leukemias. Eur J Histochem 41: 43–44

Sattler M, Quinnan LR, Pride YB, Gramlich JB, Chu SC, Even GC, Kraeft SK, Chen LB, Salgia R (2003) 2-Methoxyestradiol alters cell motility, migration, and adhesion. Blood 102:289–296

Sawyers CL (1999) Chronic myeloid leukemia. N Engl J Med 340:1330–1340

Schaller MD, Schaefer EM (2001) Multiple stimuli induce tyrosine phosphorylation of the Crk-binding sites of paxillin. Biochem J 360:57–66

Schiedlmeier B, Klump H, Will E, Arman-Kalcek G, Li Z, Wang Z, Rimek A, Friel J, Baum C, Ostertag W (2003) High-level ectopic HOXB4 expression confers a profound in vivo competitive growth advantage on human cord blood CD34+ cells, but impairs lymphomyeloid differentiation. Blood 101:1759–1768

Shanmugam K, Green NC, Rambaldi I, Saragovi HU, Featherstone MS (1999) PBX and MEIS as non-DNA-binding partners in trimeric complexes with HOX proteins. Mol Cell Biol 19:7577–7588

Shtutman M, Zhurinsky J, Oren M, Levina E, Ben-Ze'ev A (2002) PML is a target of beta-catenin and plakoglobin, and coactivates beta-catenin-mediated transcription. Cancer Res 65:5947–5954

Skorski T (2002) BCR/ABL regulates response to DNA damage: the role in resistance to genotoxic treatment and in genomic instability. Oncogene 21:8591–8604

Smith LG, Weissman IL, Heimfeld S (1991) Clonal analysis of hematopoietic stem-cell differentiation in vivo. Proc Natl Acad Sci USA 88:2788–2792

Spangrude GJ, Heimfeld S, Weissman IL (1988) Purification and characterization of mouse hematopoietic stem cells. Science 241:58–62

Spiegelman VS, Slaga TJ, Pagano M, Minamoto T, Ronai Z, Fuchs SY (2000) Wnt/beta-catenin signalling induces the expression of betaTrCP ubiquitin ligase receptor. Mol Cell 5(5):877–882

Stankovic T, Weber P, Stewart G, Bedenham T, Murray J, Byrd PJ, Moss PA, Taylor AM (1999) Inactivation of ataxia telangiectasia mutated gene in B-cell chronic lymphocytic leukemia. Lancet 353:26–29

Stone RM (2002) Treatment of acute myeloid leukemia: state-of-the-art and future directions. Semin Hematol 39:4–10

Taipale, J, Beachy PA (2001) The hedgehog and Wnt signaling pathways in cancer. Nature 411: 349–354

Tenen DG (2003) Disruption of differentiation in human cancer: AML shows the way. Nature Rev Cancer 3: 89–101

Tenen DG, Hromas R, Licht JD, Zhang, D-E (1997) Transcription factors, normal myeloid development, and leukemia. Blood 90:489–519.

Terskikh AV, Easterday MC, Li L, Hood L, Kornblum HI, Geschwind DH, Weissman IL (2001) From hematopoiesis to neuropoiesis: evidence of overlapping genetic programs. Proc Natl Acad Sci USA 98:7934–7939

Terskikh AV, Miyamoto T, Chang C, Diatchenko L, Weissman IL (2003) Gene expression analysis of purified hematopoietic stem cells and committed progenitors. Blood, Online publication

Terstappen LW, Huang S, Safford M, Lansdorp PM, Loken MR (1991) Sequential generations of hematopoietic colonies derived from single nonlineage-committed CD34+38- progenitor cells. Blood 77:1218–27

Thorsteinsdottir U, Sauvageau G, Hough MR, Dragowska W, Lansdorp PM, Lawrence HJ, Largman C, Humphries RK (1997) Overexpression of HOXA10 in murine hematopoietic cells perturbs both myeloid and lymphoid differentiation and leads to acute myeloid leukemia. Mol Cell Biol 17:495–505

Till JE, McCulloch EA (1961) A direct measurement of the radiation sensitivity of normal mouse bone marrow cells. Radiat Res 14:1419–1430

Traver D, Akashi K, Weissman IL, Lagasse E (1998) Mice defective in two apoptosis pathways in the myeloid lineage develop acute myeloblastic leukemia. Immunity 9:47–57

Traver D, Akashi K, Manz M, Merad M, Miyamoto T, Engleman EG, Weissman IL (2000) Development of CD8α-positive dendritic cells from a common myeloid progenitor. Science 290:2152–2154

Tsukamoto AS, Grosshedl R, Guzman RC, Parslow T, Varmus HE (1988) Expression of the int-1 gene in transgenic mice is associated with mammary gland hyperplasia and adenocarcinomas in male and female mice. Cell 55:619–625

Tsukamoto A, Weissman I, Chen B, DiGiusto D, Baum C, Hoffman R, Uchida N (1995) In: Levitt D, Mertelsmann (eds) Hematopoietic stem cells: biology and therapeutic applications, Dekker, New York, pp. 85–124

Turhan AG, Lemoine FM, Debert C, Bonnet ML, Baillou C, Picard F, Macintyre EA, Varet B (1995) Highly purified primitive hematopoietic stem cells are PML/RARα negative and generate nonclonal progenitors in acute promyelocytic leukemia. Blood 85:2154–2161

Uchida N (1992) Characterization of mouse hematopoietic stem cells. Stanford University PhD Thesis

Uchida N, Weissman IL (1992) Searching for hematopoietic stem cells: evidence that Thy-1.1lo Lin-Sca-1+ cells are the only stem cells in C57BL/Ka-thy-1.1 bone marrow. J Exp Med 175:175–184

Uchida N, Sutton RE, Friera AM, He D, Reitsma MJ, Chang WC, Veres G, Scollay R, Weissman IL (1998) HIV, but not murine leukemia virus, vectors mediate high efficiency gene transfer into freshly isolated G0/G1 human hematopoietic stem cells. Proc Natl Acad Sci USA 95: 11939–11944

Van den Berg DJ, Sharma AK, Bruno E, Hoffman R (1998) Role of members of the Wnt gene family in human hematopoiesis. Blood 92: 3189–3202

Varnum-Finney B, Xu L, Brashem-Stein C, Nourigat C, Flowers D, Bakkour S, Pear WS, Bernstein ID (2000) Pluripotent, cytokine-dependent, hematopoietic stem cells are immortalized by constitutive Notch1 signaling. Nature Med 6:1278–1281

Vaux DL, Cory S, Adams JM (1988) Bcl-2 gene promotes haemopoietic cell survival and cooperates with c-myc to immortalize pre-B cells. Nature 335: 440–442

Vaziri H, Dragowska W, Allsopp RC, Thomas TE, Harley CB, Lansdorp PM (1994) Evidence for a mitotic clock in human hematopoietic stem cells: loss of telomeric DNA with age. Proc Natl Acad Sci USA 91:9857–9860

Vernon-Wilson EF, Kee WJ, Willis AC, Barclay AN, Simmons DL, Brown MH (2000) CD47 is a ligand for rat macrophage membrane signal regulatory protein SIRP (OX41) and human SIRPalpha 1. Eur J Immunol 30:2130–2137

Wang JY (2000) Regulation of cell death by the Abl tyrosine kinase. Oncogene 19:5643–5650

Winn RA, Bremnes RM, Bemis L, Franklin WA, Miller YE, Cool C, Heasley LE (2002) gamma-Catenin expression is reduced or absent in a subset of human lung cancers and re-expression inhibits transformed cell growth. Oncogene 21:7497–7506

Wagers A, Sherwood RI, Christensen JL, Weissman IL (2002) Little evidence for developmental plasticity of adult hematopoietic stem cells. Science 297:2256–22259

Weissman IL (2000) Translating stem and progenitor cell biology to the clinic: barriers and opportunities. Science 287:1442–1446

Westendorf JJ, Yamamoto CM, Lenny N, Downing JR, Selsted ME, Hiebert SW (1998) The t(8;21) fusion product, AML-1-ETO, associates with C/EBP-alpha, inhibits C/EBP-alpha-dependent transcription, and blocks granulocytic differentiation. Mol Cell Biol 18:322–333

Willert K, Brown J, Danenberg, E, Ducan, AW, Weissman IL, Reya T, Yates JR III, Nusse R (2002) Wnt proteins; are lipid-modified and can act as stem cell growth factors. Nature 423: 448–452

Wodinsky I, Swiniarski J, Kensler CJ (1967) Growth kinetics of lymphocytic L1210 cells in vivo as determined by spleen colony assay. Cancer Chemother Rep 51:415–421

Wu AM, Siminovitch L, Till JE, McCulloch EA (1968) Evidence for a relationship between mouse hematopoietic stem cells and cells forming colonies in culture. Proc Natl Acad Sci USA 59: 1209–1215

Subject Index

Printing: Saladruck, Berlin
Binding: Stein+Lehmann, Berlin